AF386854

THE SPANISH ARMADA FROM THE SPANISH PERSPECTIVE

THE SPANISH ARMADA FROM THE SPANISH PERSPECTIVE

The Tudors' Greatest Naval Victory

J. J. HERRERO GIMÉNEZ

THE SPANISH ARMADA FROM THE SPANISH PERSPECTIVE
The Tudors' Greatest Naval Victory

First published in Great Britain in 2025
by Frontline Books
An imprint of
Pen & Sword Books Ltd
Yorkshire - Philadelphia
Copyright © J. J. Herrero Giménez
ISBN 9781036119324

Typeset by Lapiz Digital
Printed and bound in the UK by CPI Group (UK) Ltd,
Croydon, CR0 4YY.

Printed on paper from a sustainable source by
CPI Group (UK) Ltd, Croydon, CR0 4YY

The Publisher's authorised representative in the EU for product safety is
Authorised Rep Compliance Ltd., Ground Floor, 71 Lower Baggot Street,
Dublin D02 P593, Ireland.
www.arccompliance.com

For a complete list of Pen & Sword titles please contact
PEN & SWORD BOOKS LTD
47 Church Street, Barnsley, South Yorkshire, S70 2AS, England
E-mail: enquiries@pen-and-sword.co.uk
Website: www.pen-and-sword.co.uk
or
PEN & SWORD BOOKS
1950 Lawrence Rd, Havertown, PA 19083, USA
E-mail: uspen-and-sword@casematepublishers.com

CONTENTS

LIST OF PLATES

LIST OF MAPS AND GRAPHICS

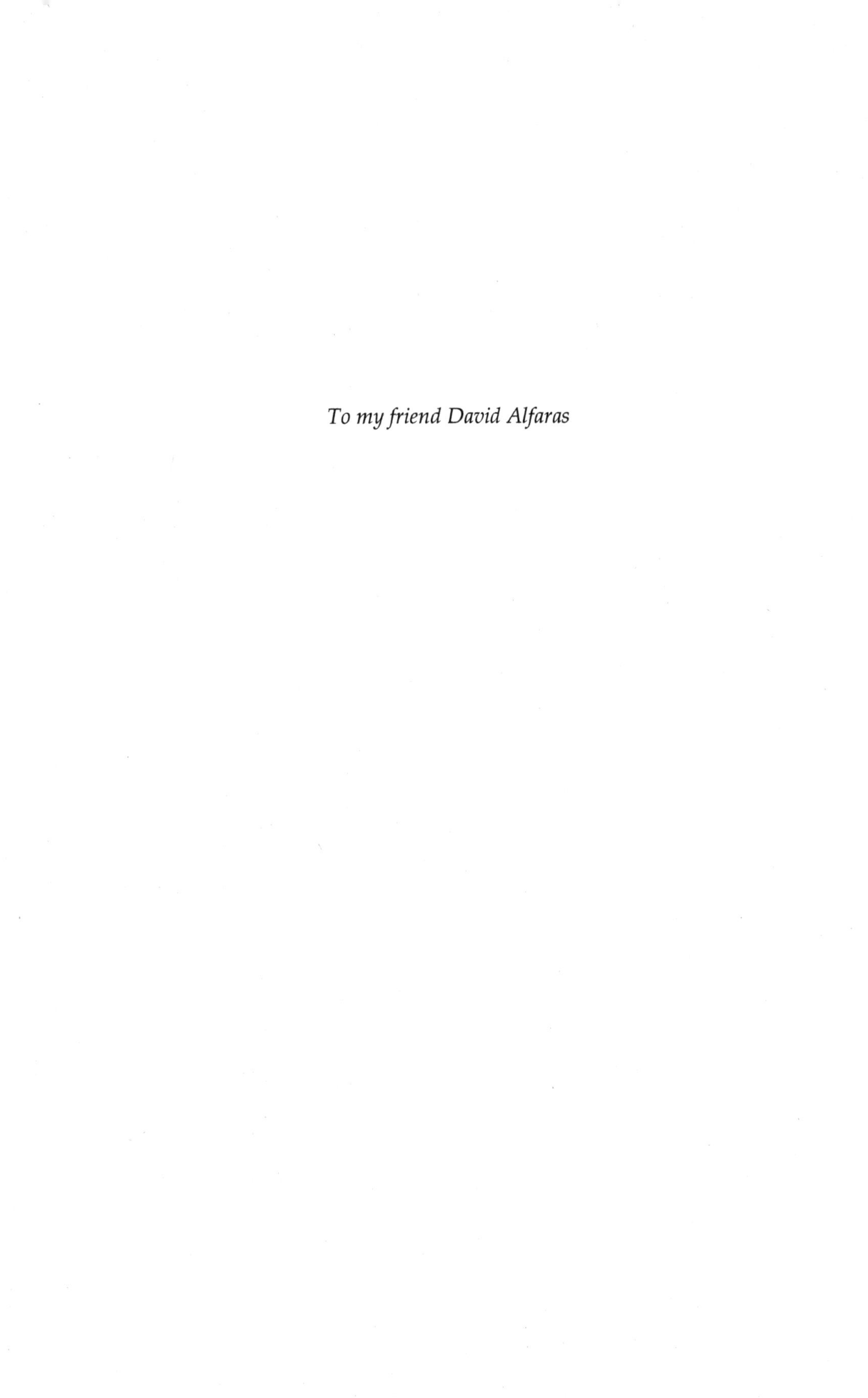

To my friend David Alfaras

ACKNOWLEDGMENTS

Although writing is an individual task, this book would not have been possible without the collaboration and help of many people. To begin with, I would like to thank John Grehan and all the team at Pen & Sword for their continued confidence and for allowing me to bring aspects of Spanish history to the English-language reader. This is the third book I have written for them, after *The Peninsular War* and *The Conquest of Iberia*, although I recognise that the subject matter of this book has been a challenge, as I have had to analyse an episode that not only concerns the history of Spain but is especially important in the history of the United Kingdom.

Impossible to forget José León and Mamen Otero. Their support is as constant and unwavering as the waves of the sea lapping against the shore. Likewise, I cannot forget my good friend Jaume Boguñà, a scholar of history, who read the first manuscript and who gave me ideas and reflections that form part of this book. From a more technical point of view, I must thank the help and patience of Rosa Busquets of the Museu Marítim de Barcelona and César of the Cathedral of Sigüenza, as well as Anne van den Dool of the Museum of Lakenhal.

For some years now, we in Spain have been enjoying a new batch of historians who are publishing excellent monographs on the Spanish naval forces. For this book I have relied on the oceanic generosity and patience of Pedro Luis Chinchilla and, most especially, of Antonio Luis Gómez Beltrán. Of the former, I would like to highlight his exhaustive work on the shipwrecks of the Armada, while of the latter, his technical studies on ships, cannons and combat tactics are simply revolutionary.

Finally, I cannot forget my family. Time is a limited commodity and the time I devote to research or writing cannot be devoted to them. I must also admit that my children, Ferran and Valèria, now teenagers, have probably not even realised that I have spent the last two years working on this book, although I hope that in the future they will consider it part of my legacy. My patient wife, Yolanda, has been much more aware of this. To all of them, most sincerely: thank you very much.

J.J. Herrero Giménez
17 March 2025

INTRODUCTION

The defeat of the Armada in 1588 marked a turning point in history, marking the decline of the Spanish Empire and the beginning of the global dominance of the Royal Navy under the bold and ruthless policies of Elizabeth I. This commonly accepted premise has been explained and justified in dozens of books and documentaries, in which numerous English historians have endeavoured to find an explanation for the fact that a small nation, England, with scarcely any economic means, was able to overcome the greatest naval power the world had ever known. In short, their conclusion is that this turning point is based on the fact that the English built better galleons (faster, more manoeuvrable), made better cannons and had better sailors and gunners than the Spanish. Thus, they defeated the Armada and, if I may *reductio ad absurdum*, England became an international power. This premise, accepted for many decades even by Spanish historiography, raises a series of questions whose answers call into question this deterministic view of the events that took place in the English Channel in 1588. Thus, in this book, in addition to explaining the planning, development and return of the Armada, I will answer some questions that deserve to be revisited: was there really an English superiority in the quality of the ships, crews, cannons and battle tactics? Did the defeat of the Armada mark the end of Spanish supremacy in the Atlantic? And, finally, did it start what has come to be known as England's Golden Age? The first attempt to try to explain the battle from Spain was made in 1885, with the work of Fernández Duro, whose monumental work of study and excavation of Castilian archives is still a reference work for historians today, although some of the information he provided has already been shown to be mistaken. Thus, we must set 1988 as the year when Spanish historiography ceased to uncritically assume the English version of the battle. But that year, coinciding with the commemoration of the 400th anniversary of the battle, young historians such as Casado Soto and Hugo O'Donnell began to publish books and articles questioning what we might call the traditional version of the events. However, it was not until well into the twenty-first century that a new batch of historians managed to get the debate out of the classrooms of the history faculties of Spanish universities and published a series of books aimed at the non-specialist public that no longer accept the traditional version of the Battle of Gravelines or the final fate of the Armada's ships, whose main exponents have been the works of Gorrochategui, Gómez Beltrán and Chinchilla, among others, the results of which have, if I may use the metaphor, became a turning point in the study of the Armada of 1588 and the consequences of its defeat. And while it is no less true that there is still much work to be done, the Anglo-Saxon public's view of the Armada of 1588 is still more myth than reality, despite a broader view of the subject by historians such as Geoffrey Parker or a slightly lighter view (because the format did not allow otherwise), such as that of Lucy Worsley in the entertaining and interesting BBC series entitled *Royal History's Biggest Fibs*, an episode of which she dedicated to the Armada. Therefore, with this book, dear reader, I do not intend to invade England (let's face it, that is something we Spaniards are not very good at), but I do intend to provide the Anglo-Saxon reader with a different view of the events of the summer of 1588. Evidently, from the most rigorous and serious point of view. History, understood as a social science, cannot be a mouthpiece for nationalist proclamations, nor can it be a place to make value judgements on events that took place more than 400 years ago. Let us settle the debate at the outset: in this story there are neither good guys nor bad guys; nor will we find the world divided into black and white. This is the story of two nations who fought for their primacy in the world. And they used every means at their disposal. Thus, the aim of this book is to offer the English-language reader this new version of the facts. New answers that I believe explain the failure of the Armada by demystifying some assumptions hitherto considered immutable. Indeed, one of the first myths to be debunked in Spain itself is the name given to the fleet of ships that attacked the English coast in 1588. While in the anglosphere it is known as the Spanish Armada, curiously in Spain it is still called the *Invincible* Armada. This derisive and ironic adjective to the point of humiliation has remained stuck like a thorn or, dare I say it, like a bull's horn, in Spanish pride over the centuries, being used to remind us of the 'perfidy' of the English. But it has also become so embedded in the Spanish collective imagination that historical works related to this event, and which refute the traditional version of the Armada expedition . . . continue to use it! This adjective has usually been attributed to Sir William Cecil, one of Elizabeth I's principal advisors, who wrote a document entitled *The copie of a letter sent out of England to Don Bernardin Mendoza ambassadour in France for the King of Spaine* and spread it throughout Europe as an art of political propaganda developed by Elizabeth I. However, in the original version of the document we do not find the adjective 'Invincible', which we do find in the Italian translation by Petruccio Ubaldini, a soldier, translator and calligrapher in the service of the English crown. But this will be the first of many myths or inaccuracies that I intend to explain and in which we will have to pillory one of the founding myths of the English nation and, by extension, of Elizabeth I's reign.

Therefore, with the research carried out by Spanish historians and my own, in this book I will analyse this battle in the following chapters. First, we will see the historical context prior to the war in Chapter 1,'Adam's Will and Testament'. In Chapter 2, 'The Swell before the Storm', we will evaluate the economic, political and religious issues that led to the war. In Chapter 3., entitled '1585–1588: the Early Battles', we will look at the main battles that took place before the campaign of the Armada in 1588, which we will explore in more detail in Chapter 4, 'The Enterprise of England', when we will explain the fighting that culminated in the Battle of Gravelines and the painful return of the surviving ships to Spain. Chapter 5, 'Anatomy of a Battle' is probably the most important chapter of this book, as it will provide an in-depth analysis of what happened at Gravelines and will be essential for understanding the true military significance of the battle. No less important is Chapter 6, 'Beyond 1588', because one of the serious flaws in understanding the battle is that it has been taken out of its military context: there was a war, and that war did not end in the summer of 1588. In fact, it lasted 16 more years. The end of the war and the Treaty of London is discussed in Chapter 7, 'A Rainy Day in Valladolid'. We will end in a chapter devoted to the conclusions, which I hope will give you a new vision of what happened in the waters of the English Channel, which, dyed red, bore witness to an epic battle between two nations fighting for world hegemony and which wrote for their histories an episode full of courage and heroism. And I cannot forget a postscript that I hope will surprise you and bring a smile to your face.

But before starting with the book, it is necessary to answer a question: was the Battle of Gravelines the Tudor's Greatest Naval Battle? And what other naval battles could have this honour? In terms of the number of ships, it was second only to the English Armada of 1589, but the latter was not involved in any major naval battles. A major engagement was the Battle of the Solent, which took place in 1545, during the reign of Henry VIII. On this occasion, the French fleet was superior to the

Spanish fleet, but the ships were smaller in tonnage. It must be said that on this occasion the invasion also failed. From a military point of view, the consequences were not very different from those of the Spanish Armada of 1588. However, the battle that took place during the reign of Henry VIII has been largely forgotten, whereas the battle of 1588 is considered to be one of the most momentous in English history. In the course of this book we will see why.

Chapter 1

ADAM'S WILL AND TESTAMENT

The origin of this story goes back several decades before the start of the war between England and Spain. Specifically, we could set it in March 1493, after Christopher Columbus returned from his first voyage to America. When his first discoveries became known, the kingdoms of Portugal and Castile disputed the ownership of the lands discovered and the ownership of their riches. On this occasion, the diplomatic route prevailed over the military one and the two Atlantic powers, under the authority of Pope Alexander VI, who incidentally was Valencian, shared the discovered territories, those to be discovered in the future and, surprisingly, even the ownership of the Atlantic Ocean, through the Treaty of Tordesillas (1494). It was the first time, but not the last, that the greed and arrogance of European countries divided up the world as they pleased.

Obviously, England, France and the Dutch Republic did not recognise the validity of this treaty. Years later, an angry Francis I of France would ask 'to see the clause in Adam's will and testament that excludes me from sharing the world'. The first step of the English and French was to organise their own expeditions to discover new passages to Asia, since at that time it was not yet realised that Columbus had in fact reached a new continent. But both failed. In the English case, John Cabot's expeditions (1496, 1497 and 1498) did not result in the establishment of a colony. It was not until the reign of Elizabeth I, in 1585, that the first stable English settlement was established, known as Roanoke Colony (in modern North Carolina, USA), with the Queen's intention of consolidating a base from which to attack Spanish bases in the Caribbean. The new settlement had approximately 120 colonists, but virtually nothing is known of their fate. In fact, when an

English ship arrived in 1590, it found no trace of those settlers. What happened to them is unknown: whether they were massacred by the Indians or integrated with them, died of starvation, moved on to another settlement or were annihilated by the Spanish. In the case of France, it was the navigator Binot Paulmier de Gonneville, who tried unsuccessfully to establish colonies in Brazil. These failures led their respective countries to a second phase: if they were unable to establish themselves in America by their own means, they would try to seize the riches of the New World through the action of privateers. Thus, France, Spain's bitter enemy during the first half of the sixteenth century, soon sent privateering expeditions to America, notably those of Jean Fleury and Jacques de Sores. The success of some of their expeditions made known throughout Europe the formidable quantities of gold and silver that the Spanish were extracting from the Americas. England, for its part, kept out of the corsair activity, as it had been a staunch ally of Spain since the signing of the Treaty of Medina del Campo in 1489,[1] at a time when it seemed that the alliance between the two countries was a natural one. On one hand, Henry VII gained an ally regarding his territorial claims against France (since England had not renounced recovering Brittany) and managed to internationally legitimise the new dynasty he inaugurated after the Wars of the Roses.[2] Moreover, along with his marriage to Elizabeth of York, he secured a strong external ally in the face of potential internal opposition. As for Spain, the agreement was full of advantages. Ferdinand II of Aragon had skilfully woven a network of alliances with the countries surrounding France. First with the Habsburgs, and now with the Tudors. The agreement was sealed with the promise of marriage between Arthur, Prince of Wales, and Catherine of Aragon, the youngest daughter of the monarchs of Spain. However, since the future spouses were only three and two years old respectively, the wedding did not take place until 1501. The following year, Arthur died, and both Edward and Ferdinand agreed that Catherine would marry the new Prince of Wales, the future Henry VIII.

At this point, it is worth noting how the diplomatic policy implemented by Ferdinand II of Aragon succeeded in forging a coalition system through dynastic marriages – not only with England, but also with Portugal, via the House of Avís, and with the Austrian House of Habsburg, from which the problematic territories of the Low Countries would later be inherited. It may now be appropriate to highlight a key aspect of Ferdinand's worldview that would influence subsequent Spanish monarchs and, consequently, shape the future conflict between England and Spain: namely, the central importance that the King of Aragon attributed to his power and influence in the Mediterranean. To properly understand this, one must bear in mind that Spain, at the time, did not exist as a unified nation-state in administrative terms. Rather, it functioned as a confederation of the Kingdoms of Castile and Aragon, dynastically united through the marriage of Isabella of Castile and Ferdinand of Aragon.[3] The Catholic Monarchs shared governmental responsibilities equally, though at times the strategic interests of their respective kingdoms were at odds – often due to geographic determinants. Historically, Castile had oriented itself toward the Atlantic, with France often regarded as a natural ally. In contrast, Aragon had built an imperial presence in the Mediterranean and viewed France as a long-standing mortal enemy, one with which it had contested control over territories north of the Pyrenees and in Italy for centuries. Within this Mediterranean-oriented cosmovision, the principal adversaries were France and the Ottoman Empire. In this evolving web of alliances, England came to be seen as a preferred and strategic partner of Spain.

Let us move forward in history to the year 1509. In England, Henry VIII ascended to the throne at the age of 18. The young Tudor, eager to reclaim part of the territory lost by the Plantagenets during the Hundred Years War, enthusiastically joined the Holy League proclaimed by Pope Julius II in 1511 to counter French expansionism. This alliance included the Papal States, Venice, the Holy Roman Empire, the Swiss Confederacy, England, and Spain. Only a few weeks later, Henry VIII and Ferdinand II signed the Treaty of Westminster, by which both monarchs committed to mutual military support against France. Immediately thereafter, Henry VIII and Ferdinand II began preparations for the invasion of Gascony. In May 1512, the English king dispatched a force of 10,000 soldiers to Irún (near the French border), under the command of Thomas Grey, Marquis of Dorset. These troops were to join the Spanish forces, composed of an additional 6,000 infantrymen, 2,500 cavalry and siege artillery, all under the command of Fadrique Álvarez de Toledo, Duke of Alba.[4] However, Ferdinand's priority was not to support the English, but rather to conquer the Kingdom of Navarre.[5] Thus, it could be said that, in one of his characteristic stratagems, Ferdinand used the English army in July to threaten France, while the Duke of Alba's army proceeded to conquer Navarre, which was unable to receive support from its allies. However, the prolonged inaction began to take its toll on the English camp, where dysentery broke out, followed by a mutiny due to delayed payments. In August, the English troops sacked Irún, and the French, realising the disorder within the enemy ranks, decided to launch an attack. Forewarned by spies, Dorset pre-emptively moved and succeeded in capturing the French town of Saint-Jean-de-Luz, even reaching as far as Bayonne. Yet, the lack of artillery and siege equipment forced him to retreat to Irún. In September, Dorset resolved to return to England, but Ferdinand promised to dispatch his army immediately to begin the campaign, sending 400 cavalrymen ahead as an advance guard. Nevertheless, several of Dorset's captains mutinied once again, and as a new epidemic spread through the camp, the English general fell ill with dysentery. He was eventually replaced by Thomas Howard, 3rd Duke of Norfolk.[6] Due to Ferdinand's lack of cooperation, letters were sent to London requesting the return of the troops; however, Henry VIII ordered that his forces remain in Spain until the spring. Nevertheless, Howard and his captains, weary of the unfulfilled promises made by their supposed allies, took the initiative to return to England on their own accord. Upon their arrival in London, it appears that Henry VIII reacted with great anger and ordered an investigation into the events. However, it seems that his anger was directed more at his father-in-law than at his own soldiers, who were not punished. This may have been one of the reasons behind his decision to sign a peace treaty with France, which was sealed by the marriage of his sister, Mary Tudor, to King Louis XII of France. This marriage, however, lasted only three months due to the French king's death in 1515. His successor, Francis I, eager to confront Spain, sought to reach an agreement with Henry VIII in June 1520 during the renowned summit known as the Field of the Cloth of Gold. The negotiations, however, failed – partly due to the influence of Catherine of Aragon, still the English king's wife at the time. In contrast, Charles I of Spain successfully renewed his alliance with England through the Treaty of Windsor. Thus, cordial relations between Spain and England persisted, even during the crisis that arose when Henry VIII resolved to divorce Catherine of Aragon and break away from papal authority, thereby separating from the Church of Rome. In any case, Charles I of Spain was engaged on multiple fronts across Europe and continued to prefer England as an ally rather than opening yet another theatre of war. As a point of reference, during those same months, Charles was in Vienna assisting his brother Ferdinand in resisting the siege imposed on the city by Suleiman the Magnificent – successfully halting the Ottoman Empire's advance over land.

The brief reign of Edward VI, despite deepening the break with Rome and the intrigues of the Duke of Northumberland, did not result in a significant shift in England's policy of friendship toward Spain, although diplomatic relations cooled somewhat. Following the death of the young king and the brief interlude of Lady Jane

Grey, Mary Tudor was proclaimed queen, thus restoring England to Catholic obedience. Like her grandfather Henry VII, Mary understood the necessity of securing a strong ally to consolidate her power, and no candidate seemed more suitable than the eldest son of Charles I of Spain – her cousin, Philip II – whom she married at Winchester Cathedral on 25 July 1554. It appears that Mary genuinely fell in love with the Spanish prince. Philip, then 26 years old, was described as handsome, refined and meticulously groomed. However, it seems his feelings were not reciprocated. Mary, at 38, was reportedly unattractive, toothless (due in part to an excessive fondness for sweets), and careless in her personal appearance. Regardless, Philip regarded kingship as a duty, one of whose principal responsibilities was to produce an heir. To that end, he summoned a notary to formally attest to the consummation of the marriage. Mary soon believed herself to be pregnant. Physicians predicted that the child would be born around mid-June 1555. However, no signs of labour emerged, and by early July, it became clear that the pregnancy had been a false one – a case of phantom pregnancy. The emotional and political impact on Mary was devastating, both as queen and as a woman. Her state of mind worsened when, just weeks later, Philip was summoned by his father to the Low Countries. Charles I of Spain was preparing to abdicate. He divided his vast territories between his brother Ferdinand, Archduke of Austria – who received the Holy Roman Empire – and Philip, who was granted the kingdoms of Sicily, Naples, and Sardinia, the Low Countries, the Duchy of Burgundy, and ultimately the Spanish realms of Castile and Aragon, including, of course, their American possessions. This was an extraordinary inheritance, but one that came with a heavy burden. With these territories, Philip II also inherited numerous and costly wars. Indeed, one of his first major decisions was to declare a state bankruptcy in April 1557.

Within the context of the interminable Italian Wars, the French once again launched an attack on the Kingdom of Naples. Philip returned to England to request assistance from his wife; however, as the terms of their marriage did not compel the English to participate in Spanish military campaigns, Philip was met with a refusal from the English Parliament. Shortly thereafter, Thomas Stafford, arriving from France, seized Scarborough Castle and incited a rebellion against the Catholic Queen Mary. Although the uprising was swiftly crushed and Stafford executed, the incident reignited fears regarding the expansionist policies of Henry II of France. Consequently, it was agreed that war would be declared on France (on 7 June 1557), and a force of 5,000 infantry and 1,000 cavalry would be dispatched under the command of the chameleon-like William Herbert,[7] the Earl of Pembroke. This contingent arrived too late to take part in the Battle of Saint-Quentin, which proved to be a decisive Spanish victory. However, Herbert's men fought valiantly in the subsequent siege of the city, which was soon taken. Thereafter, English and Spanish forces joined in the engagements at Le Catelet, Noyon, and Ham. During this same war – specifically on 8 January 1558 – French troops under the Duke of Guise captured Calais, thereby marking the definitive loss of England's last remaining possession on the European continent. That same year, on 17 November, Mary Tudor died, and Elizabeth I was proclaimed queen. Philip, who can hardly be described as a grieving widower, proposed marriage to the new queen, but his offer was declined. It is worth noting that the two had always maintained a cordial relationship, and indeed, it is documented that Philip had previously defended Elizabeth before her sister Mary. The following year saw the signing of the Treaty of Cateau-Cambrésis – arguably the most significant diplomatic agreement of the sixteenth century – which secured peace between Spain and France for nearly a century and confirmed the supremacy of the Spanish monarchy in European affairs. The treaty was sealed through the marriage of Philip II to Elisabeth of Valois, the eldest daughter of King Henry II of France, who, incidentally, died accidentally during the wedding celebrations. While participating in a tournament, he was fatally wounded in the eye by a splinter from the lance of Gabriel de Lorges, Count of Montgomery,[8] which caused septicaemia and, subsequently, his death. It can be said that the marriage to Elisabeth of Valois was the happiest of the four unions Philip celebrated. The French princess, only 14 years old, lively and spirited, brought a sense of vibrancy to the Royal Palace. She was nothing like his two previous wives, and it can be affirmed that, for the first time, Philip II was truly and happily in love.

Chapter 2

THE SWELL BEFORE THE STORM

'Whoever controls trade will control gold, and whoever controls gold will control the world.' This phrase, or one similar to it, attributed to Sir Walter Raleigh, likely inspired Elizabeth I to develop a policy that could be described as anti-Spanish. Her objective was to weaken Philip II's government and expand England beyond its island borders, either by exerting influence on the European continent or by colonising and trading with the Americas. Throughout her reign, Elizabeth gradually took steps to confront Spain. Initially, these actions were minor provocations that King Philip preferred to ignore. However, as the situation escalated, he eventually had to address what is commonly referred to as 'the elephant in the room'. The so-called 'Prudent King' had run out of patience and decided to confront the woman he referred to as his dear sister. In the following section, we will briefly analyse the primary areas in which this conflict unfolded: economic, religious, and territorial, the latter encompassing ramifications in Portugal, Flanders, Morocco and the Ottoman Empire.

The Sea Dogs, Ships Laden with Gold, and the Economic Question

The Treaty of Tordesillas granted Spain and Portugal a monopoly on trade with the New World. Faced with this restriction, other European nations had only two options: accept it passively or defy it by confronting the two dominant powers. England was the first nation to rebel against this perceived injustice, funding two expeditions led by John Cabot during the reign of Edward VII, though these ventures yielded no significant results. Years later, France followed suit when its corsair Jean Fleury captured part of the Aztec gold in 1522. This event revealed to Europe the immense wealth hidden in the Americas and ignited the imagination and greed of monarchs and adventurers across the continent. This marked the beginning of the first phase of piracy in the Caribbean, characterised by Spain's weak defensive system. Spanish cities and ports storing treasures from the New World became frequent targets of these attacks, which intensified throughout the century. Similarly, many merchant ships were assaulted with impunity. The situation grew so dire that Charles I and later Philip II allocated substantial resources to defend their shipments of gold and silver. The primary measure taken was the development of a protection system for merchant fleets, which resulted in the establishment of a convoy system known as the Spanish Treasure Fleet. Although the system began to function tentatively after Fleury's attack in 1522, its definitive development was carried out by Pedro Menéndez de Avilés a few years later, establishing it as one of the most successful trade routes in history. For instance, historian Pierre Chaunu estimates that between 1540 and 1650, out of 11,000 vessels, only 107 were lost due to pirate or corsair attacks. Even so, each case should be examined individually, as many of these ships risked travelling alone, unable to wait for escort by the West Indies Fleet. Regarding fortifications, Philip II sent the engineer Battista Antonelli to America. Antonelli had already effectively fortified Spain's Mediterranean coast and was tasked with designing defences for ports such as Cartagena de Indias, Havana, Veracruz (San Juan de Ulúa), and Portobello. However, financial constraints delayed construction in many cases. During these early years, French corsairs frequently targeted Spanish bases in the Caribbean. At that time in England, Henry VIII had signed the Treaty of Windsor with Charles I of Spain, preventing the Tudor king from antagonising his powerful ally with a 'corsair adventure'. Nevertheless, he pursued alternative routes to reach the East Indies. These attempts included John Rut's westward expedition in 1527–8 and Sir Hugh Willoughby's eastward venture; both ended in failure. Toward the end of his reign, Henry Tudor sent Martin Frobisher on two expeditions to West Africa, which were commercially successful. However, during his second voyage, the intrepid English sailor was imprisoned by the Portuguese for four years.

It was not until Elizabeth's reign that the situation changed radically. This shift in approach occurred in the early 1560s, when John Hawkins managed to break the monopoly on trade with the Americas through unorthodox methods. English ships stationed themselves near Caribbean settlements where they sought to trade, typically involving the sale of slaves[1] and English manufactured goods. If Hawkins encountered resistance to negotiation, he would threaten to bombard the town, carrying out his threat and subsequently looting it if the refusal persisted.[2] His first two expeditions proved highly profitable, but the third, launched in 1567, ended in disaster. Accompanied by his cousin Francis Drake, Hawkins led an expedition consisting of six ships. The journey began positively, but various setbacks damaged several vessels in the fleet. Hawkins had no choice but to approach the port of San Juan de Ulúa (modern-day Veracruz, Mexico) for repairs. He negotiated with local authorities to repair the ships and be allowed to leave peacefully in exchange for refraining from attacking the town. However, days later, a fleet of thirteen ships (one war galleon and twelve armed transport vessels) under Admiral Francisco de Luján arrived at San Juan de Ulúa, escorting the new Viceroy of New Spain, Martín Enríquez de Almansa. Hawkins ordered a Spanish officer from the port to urgently approach the newly arrived fleet to explain the non-aggression agreements he had reached. Apparently, the new viceroy agreed and accepted these terms. However, Admiral Francisco de Luján declared that he would not negotiate with corsairs and decided to take action, launching an attack on the unsuspecting English forces. The date was 24 September 1567. One of Hawkins' ships was sunk, and three others surrendered. Only the *Judith* and the *Minion* managed to escape, leaving many of their compatriots stranded on land.[3] After a harrowing journey, the two vessels managed to dock at an English port on 20 January 1568.[4] Upon his return, much like Hannibal swearing eternal hatred for Rome, Drake vowed eternal hatred for Spain and dedicated his life to fiercely combating it – literally until his final day. He did not take long to seek revenge. Little is known about Drake's expeditions in 1570 and 1571. His first major success came during the expedition of 1572–3, when he departed England with two ships, the *Pascha* and the *Swan*, crewed by seventy-three sailors, including his brothers John and Joseph.[5] The primary objective of the mission was to seize Nombre de Dios, located on the Isthmus of Panama, to serve as a base for future corsair expeditions. However, the attack failed, and Drake himself was wounded in the leg by a musket shot. Following this setback, he set course for Cartagena de Indias, where he managed to capture a merchant vessel. It is presumed that, while replenishing supplies along the coast, Drake established contact with a group of Maroons,[6] who were in constant conflict with Spanish authorities. Many of these Maroons had previously been slaves transporting gold and silver via mule trains from Peru to Nombre de Dios, where the treasure was shipped to Spain. The Maroons informed Drake of the proximity of a mule convoy carrying royal treasure, prompting him to organise an attack. The first attempt failed. Providentially, French pirate Guillaume Le Testu appeared in the area, and they agreed to jointly assault the mule convoy. On this occasion, the corsairs succeeded in capturing the loot, with Drake's share amounting to approximately 28,500 pounds of silver.[7] However, the arrival of Spanish troops forced Drake to bury a large portion of the loot and take only what could be carried in small boats.[8] With the remaining treasure loaded, Drake and what was left of his crew set sail for London, where they were received as true heroes. Despite the hardships endured and the losses suffered due to combat and disease, they returned with a substantial fortune. Nevertheless, this would not be Drake's greatest success. That came during his circumnavigation of the globe, undertaken between 1577 and 1580.[9] Drake departed from Plymouth with four ships and a pinnace, carrying a total crew of approximately 170 men. His initial objective was to attack the undefended Pacific coast. However, navigating the Strait of Magellan proved challenging, and only the *Golden Hind* managed to cross it successfully. Along the Chilean coast, Drake attacked various settlements and seized a merchant ship carrying 25,000 pesos in gold and a spectacular emerald cross, which he later gifted to Queen Elizabeth upon his return. Yet, if any city resisted him, such as Copiapó or El Callao, Drake did not waste time in prolonged battles and continued his route. As Fray Pedro Simón aptly described him: 'He did not come to win honour but wealth.'[10] In March 1579, Drake successfully captured the galleon *Nuestra Señora de la Concepción*, which was transporting 400,000 pesos in gold – the largest prize ever seized by any English privateer of his generation. He continued sailing, attacking other vessels, until he reached Acapulco. However, finding the city well-fortified, he refrained from launching an assault. Drake then arrived at the coast of present-day California, which he named 'New Albion'. From there, he crossed the Pacific Ocean, reaching the Marianas, the Moluccas and the Philippines. Finally, in 1580, he returned to Plymouth, where he was received as a hero and knighted by Queen Elizabeth. The expedition yielded profits of approximately £250,000.[11] A star had been born. After spending several years enjoying his wealth on land, Drake organised what could be considered the first expedition of the Anglo-Spanish War in 1585, a campaign that will be discussed further in subsequent sections.

Francis Drake is widely regarded as the most significant and successful privateer for the Queen, yet he was not the sole operative of his kind. A plethora of individuals emulated the tactics employed by Drake and John Hawkins in the Spanish Main, which comprised a form of 'hit and run' strategy. This approach targeted vessels sailing unaccompanied and/or unarmed, designated as *'zorreros'*, and settlements with minimal fortifications. The spoils from these expeditions were not limited to gold, silver and precious stones. Among the most sought-after items were sugar, wine, tobacco, silks and spices. However, not all of them experienced the same fate. Some, less well known, failed in their attempt to emulate Drake. One such individual was John Oxenham, a former lieutenant under Drake who had accompanied him on his 1572–3 expedition. In 1576, Oxenham led his own expedition to Central America in command of a ship and fifty-seven sailors. He was the first Englishman to explore the Isthmus of Panama. Oxenham's endeavour was met with considerable success, yielding a substantial quantity of gold and silver. However, his triumph was short-lived, as he was captured by the Spanish and subsequently executed. The 1576 expedition, organised by Andrew Barker and William Coxe aboard the *Ragged Staffe* and the *Bear* respectively, also met with failure. In Honduras, they were ambushed by the Spanish, resulting in the death of Barker and most of his crew. Coxe, finding himself in danger, weighed anchor and sailed back to England, where he was arrested, tried and imprisoned.[12] It is worth noting that he later commanded one of the ships that engaged the Armada and was killed in the fighting.

However, if the Sea Dogs had been effective in their privateering activities, they had failed miserably in their other role as explorers and settlers. In this regard, four notable individuals merit consideration: Martin Frobisher, Humphrey Gilbert, Walter Raleigh and James Alday. Frobisher and Gilbert's obsession was to find the North-west Passage by which to reach the East Indies without having to cross the very dangerous Strait of Magellan. In 1576, Frobisher embarked on his inaugural expedition, but after a punishing voyage, he returned with only an Inuit prisoner and a worthless marcasite stone. In 1577, he successfully secured additional funding for a subsequent expedition, this time with the financial backing of the Queen herself. However, this endeavour also proved unsuccessful, yielding only pyrite, erroneously identified as gold, and three hostages who perished shortly after reaching London. Despite the apparent failure of the endeavour, however, there remained a degree of confidence in the economic potential of the territories explored by Frobisher. Consequently, in 1578, he was able to organise a new expedition, this time the largest of all, comprising fifteen ships. The voyage was once again hindered by inclement weather and internal strife between Frobisher and several of his ship captains and pilots. The voyage culminated in the discovery of a bay that would later be named Frobisher Bay, located within the contemporary Qikiqtaaluk Region of Nunavut, Canada. Despite the initial optimism, the attempt to establish a permanent settlement in this location proved unsuccessful. However, the prevailing discontent amongst the members of the expedition, compounded by the scarcity of adequate materials for establishing a viable settlement in the region, ultimately led to the decision to return to England. In 1583, Sir Humphrey Gilbert, a renowned military commander and explorer of the Elizabethan era, embarked on another attempt. Departing England in June 1583, he led a fleet of five ships. Reaching Newfoundland in August, he assumed control of the region on behalf of the Queen. However, his efforts to find a passage to Asia met with failure. In September, he was compelled to return, bearing scant supplies and confronted with significant challenges concerning discipline on his ships. A few days later, his ship the *Squirrel* sank near the Azores, and Gilbert and the rest of the sailors were lost.[13] The second of the explorers, Walter Raleigh, is another of the great seafarers of his generation. In 1584, he received permission from the Queen to explore and establish a settlement on a land 'not actually possessed of any Christian Prince or inhabited by Christian People'.[14] The settlement's objectives were twofold: firstly, to establish and strengthen a settlement in the present-day United States for trade with the indigenous population and as an operational base for attacks on Hispanic possessions in the Caribbean and the Indies Fleet. Although it was Raleigh who received the patent for this venture, he commissioned two of his servants, Philip Amadas and Arthur Barlowe, to undertake the first voyage of exploration. Departing England at the close of April 1584, the exploratory party arrived at Roanoke Island in the middle of July, where they established contact with the local Secotan tribe, with whom they maintained cordial relations. The group considered the location to be a suitable site for future habitation, and thus, a month later, they embarked on their return journey to England.[15] The following year, a new expedition led by the famous Richard Grenville[16] left 107 settlers at Roanoke under the command of Ralph Lane. The next few months were not easy, and relations with the natives became so strained that they were attacked. At the end of June 1586, Sir Francis Drake landed on the island from his last expedition to the Caribbean and agreed with Lane to ensure their return to England and abandon the colony. Another attempt to consolidate the settlement was made the following year. This time Raleigh put John White in charge. His intention was to establish a new colony on the Chesapeake Bay, but the pilots of his ships refused to go beyond Roanoke, where they left about 110 settlers. Six weeks later, White and his ships returned to England with a cargo of supplies and new settlers, but it took him three years to return to Roanoke. When he arrived, he found no trace of the settlers,[17] including his daughter and son-in-law. In short, all attempts at American colonisation during the Elizabethan period had been a complete failure.

These operations included the Raid to Newfoundland, which took place in the latter half of 1585. The objective of this raid was to expel the defenceless Spanish-Portuguese fishing vessels from a region that could be considered a kind of no man's land, where English warships would not have to face Spanish galleons. At this juncture, it is noteworthy to observe the significant emphasis that the Tudors placed on fisheries policy, evidenced by a succession of pivotal reforms during their reign. Elizabeth was an active proponent of these reforms, and in 1563, she promulgated the Act to which certain political constitutions were made for the maintenance of the navy, which consolidated various initiatives to promote the fishing industry, including the curious obligation to consume fish on Wednesdays, Fridays and Saturdays. The overarching objective of these reforms was twofold: firstly, to enhance the competitiveness of the English shipping industry and secondly, to secure fishing grounds for English vessels, which were significantly disadvantaged by the competition from other fishing powers. The North Sea was dominated by Scandinavian and Russian vessels, the Icelandic fishing grounds by Scandinavian and Dutch vessels, Newfoundland by Spanish and French vessels, and a substantial proportion of the fish arriving in England was transported in Dutch ships.[18] This issue was of particular concern to Elizabeth and her advisors, who recognised that the problem would not only impact the English economy, but also national defence, given the role of ships in defending the nation.[19] It was within this context that Drake's expedition set sail, with the objective of establishing English dominion over Newfoundland. In September, Drake assumed command of ten ships and accompanied by additional English vessels, engaged in the plundering and sinking of an unspecified number of Spanish-Portuguese fishing boats in Saint John. With regard to the spoils of war, there is some discrepancy amongst sources,[20] but it is generally accepted that Drake captured approximately 20 Portuguese fishing vessels, approximately 600 Spanish-Portuguese prisoners, and between 50,000 and 60,000 quintals of fish, in addition to two merchant ships transporting wine, sugar, and ivory. This series of events resulted in substantial financial gains for Drake and his investors. However, Bernard Drake could not enjoy his success for long. Upon his return to England, he ordered the crew of one of the Portuguese fishing boats that had mutinied to be imprisoned in Exeter. The conditions under which the crew were left were such that they were compelled to live on the charity of their neighbours. When the trial proceedings commenced a few weeks later, the crew members were all gravely ill. Indeed, a significant proportion of those in attendance perished from the disease, including Bernard Drake himself. In conclusion, the raid led by Drake was successful in driving out the Spanish-Portuguese fishermen, whilst the French continued to fish freely. Although he failed in establishing a colony, he did succeed in freeing the territory. In 1610, under the reign of James I, he settled in Cuper's Cove (present-day Cupids, Newfoundland), laying the foundation for the future English and French colonies in the north of the American continent.

Murad III, the Mediterranean and the Islamic Question

During the sixteenth century, the establishment of alliances between Christian and Islamic countries was generally discouraged. Despite the existence of periodic trade interactions, the equilibrium of power and the dynamics of diplomatic alliances often rendered these relations somewhat turbulent. However, Elizabeth I perceived the Ottoman Empire and Morocco as a potential conduit for the establishment of new trade markets and potential allies in the event of war with Spain, a manoeuvre that Francis I of France had already executed in the first half of the century with Suleiman the Magnificent. At the time of the Queen's coronation in 1559,[21] the geopolitical landscape of the Mediterranean was characterised by the pre-eminence of two powers: the Ottoman Empire held dominion over the eastern region, while Spain exercised control over the western portion. In the eastern Mediterranean, the Ottomans were gradually expanding their territorial dominion, previously held by the Republic of Venice. In the western Mediterranean, the Treaty of Cateau-Cambrésis (1559) stipulated that France relinquish its claims to the Mediterranean and the Italian territories that were the subject of a dispute with Philip II of Spain.

However, the *Mare Nostrum* was too small for the two powers with an expansionist vocation. Consequently, during the sixteenth century, Spain and the Ottoman Empire engaged in perpetual hostilities, with Philip II's fleet attaining some of its most significant victories, along with a number of its most disastrous defeats. North African enclaves such as Oran, Algiers, Tunis, Mers El Kébir and La Goulette experienced a constant shift in control between the two powers. By the mid-sixteenth century, a shift in the balance of power appeared to favour the Ottoman side. The admirals of the Sublime Porte, such as Piali Pasha and Dragut, appeared to be unstoppable. The Ottomans engaged in widespread plundering of the southern Italian peninsula and its islands, and even the Spanish coast, with impunity, and in 1560 they recaptured the Spanish enclave of Djerba. In 1565, the Ottomans initiated a plan to conquer the island of Malta. This island represented the last significant stronghold of the Order of the Knights of St John of Jerusalem. Following the loss of the Holy Land in 1310, the Order relocated to Rhodes, where they engaged in privateering, not only against Muslim ships. However, in 1520, Suleiman the Magnificent expelled them after a six-month siege. Ten years later, Charles I of Spain, with the approval of Pope Clement VII, ceded them several of his own territories, thus providing the Order with a new base: Tripoli and the islands of Malta, Gozo and Comino. In consideration of the aforementioned cession, the King of Spain was to receive one Maltese falcon on an annual basis. Despite the fall of Tripoli to the Ottomans the following year, the Knights of St John maintained their hold in Malta and its two small neighbouring islands, continuing their privateering operations, this time focusing on the route from Egypt to Constantinople. Suleiman perceived the conquest of Malta as a strategic opportunity to quell the persistent threat posed by Christian piracy,[22] while also serving as a springboard for potential further expansion into Italy and beyond.[23] The strategic significance of this confrontation was such that it was widely regarded as having the potential to influence the course of European history. The siege commenced in mid-May 1565. The battle was marked by an apparent imbalance in combatant numbers, with 35,000 Ottomans facing 6,000 Maltese defenders, including 500 knights of the Order of St John and 400 Spanish soldiers of the Tercio Viejo de Sicilia, many of whom sought refuge in Fort Saint Elmo, which would become a kind of Alamo or Battle of Thermopylae for the Order. Following a month of relentless combat, the fort was reduced to rubble and its defenders were massacred. However, Mustafa Pasha, the Ottoman general-in-chief, had lost more than 3,000 janissaries (his elite troops) and Dragut, his most prominent commander, who perished in the fierce fighting. In the following months, the Ottomans endeavoured to breach the city walls of Burgi and Senglea, but to no avail. Notably, at the Bastion of Castile, a particularly intense combat zone, it is recorded that the defence was led by an English knight of the Order, Sir Oliver Starkley.[24] Finally, in mid-September, Mustafa Pasha gave the order to lift the siege and winter in Mdina, in the centre of the island. However, the arrival of a Spanish fleet under the command of Álvaro de Bazán and Andrea Doria, accompanied by a contingent of 10,000 troops, prompted the Ottoman general to embark his men and return to Constantinople. This development ultimately ensured the preservation of Malta. The Ottomans swiftly recuperated from this setback, expelling the Venetians from Cyprus in 1570. In response, the papacy organised a new Holy League. This new alliance, unlike its predecessor, was to bring together the two largest fleets the world had seen since the Battle of Actium (31 BC), with a total of approximately 600 ships. The Holy League Armada was under the command of John of Austria,[25] the half-brother of Philip II, a young man of 26, who possessed extraordinary gifts for war, as he would demonstrate throughout his short but eventful life.

Additionally, he would benefit from the counsel of some of the finest Spanish admirals, such as Lluís de Requesens and Álvaro de Bazán, as well as those serving the Spanish Crown, including the Genoese Gianandrea Doria (who commanded the right wing), alongside his cousin Alessandro Farnese. The left wing was under the command of the Venetian Agostino Barbarigo, while the centre was led by John of Austria himself. In a tactical arrangement like that of the Catholic forces, Ali Pasha, the Ottoman fleet's commander-in-chief, occupied the centre of the line; Mehmed Siroco's galleys faced Barbarigo's Venetians; and Occhiali's[26] galleys confronted those of Doria. The first clashes erupted on the Holy League's left wing between Barbarigo's and Siroco's galleys. The engagement was so fierce that both admirals lost their lives during the battle. In the centre, Ali Pasha faced John of Austria, where the decisive combat took place. Spanish soldiers clearly outmatched their Ottoman counterparts thanks to their tactic of sweeping the decks of enemy galleys with musket and arquebus fire – weaponry not unfamiliar to the Ottomans but less favoured by their soldiers, who predominantly relied on bows and arrows, less effective against the armour worn by Spanish troops. At the climax of the battle, the two flagship vessels – the *Sultana* of Ali Pasha and John of Austria's *Real* – engaged directly. After two failed attempts, Spanish forces successfully boarded Ali Pasha's flagship, where he perished in hand-to-hand combat. Despite desperate efforts to reverse the situation, victory was assured for the Catholic forces. Only a manoeuvre by the Genoese Doria on the right wing created a gap that Occhiali attempted to exploit to encircle John of Austria's squadron. However, Álvaro de Bazán swiftly moved to cover the gap left by Doria, crushing the only opportunity available to the Sultan's galleys. The Holy League had won, and the victory was overwhelming. Ali Pasha's fleet lost 210 galleys out of the 278 deployed in the battle, amounting to 75.5 per cent, along with a similarly significant number of soldiers and sailors killed, wounded, or captured. Was the Battle of Lepanto a decisive victory? Yes, it was. The battle tacitly established a status quo in the Mediterranean whereby the Ottoman Empire would control the eastern half, while Spain would dominate the western half – a balance that lasted until the early eighteenth century, when Philip V of Spain lost his Italian possessions following the War of the Spanish Succession (1701–15) and the Treaty of Utrecht (1713). Although both empires continued their conflict, it did so with much lower intensity. In 1573, an exhausted Venice signed a peace agreement with the *Sublime Porte*, and in 1580, Philip II agreed to a truce with Murad III. Neither side had any interest in continuing a Mediterranean war while facing open fronts elsewhere and enduring financial crises. A year before the signing of this truce, formal relations between England and the Ottoman Sultan began, culminating in the establishment of the Turkey Company in 1580,[27] which granted England preferential treatment for its merchants. William Harborne was appointed to oversee operations in Constantinople, serving simultaneously as England's ambassador to the Sultan. The creation of the company enabled England to access a wide range of goods without relying on Venetian or Dutch intermediaries. In 1583, the Venice Company was established to facilitate trade with the Republic of Venice, and both companies merged in 1592 to form the renowned Levant Company. However, the Levant Company drew complaints from many Christian nations, as English merchants exported war-related materials to Turkey – such as tin, lead and strong fabrics used for Janissary uniforms[28] – effectively making Queen Elizabeth I an adversary of Christendom. Harborne's intentions were not purely economic; he also sought to influence the Sultan's political decisions, particularly regarding Spain. For example, English agents successfully ensured that the truce with Spain was not renewed in 1587. However, this had no practical consequences, as Spain and Turkey did not engage in further conflicts despite England's persistent calls for opening a new front against Philip II. Murad III was preoccupied with his ongoing war against the Safavids and his campaigns against Habsburg territories in Central Europe. Nevertheless, relations between England and the Ottoman Empire remained remarkably cordial. This was especially true between Queen Elizabeth I and Safiye Sultan, Murad III's favourite wife and mother of Mehmed III – a woman of significant influence who admired Elizabeth for being a female ruler. Additionally, a sentiment of admiration toward the English emerged among influential Ottoman generals, such as Hasan Pasha, who respected how a small island could pose substantial challenges to a superpower like Spain.

The other Islamic geographical sphere where Elizabeth I managed to exert some influence was Morocco. If we look back to the mid-sixteenth century, the situation in the country was far from stable. The territory was contested by two dynasties, the Sa'adi and the Wattasid, with the former ultimately prevailing in 1549. The Sa'adi dynasty rose to power with the support of the Kingdom of Castile, while facing opposition from the Ottoman Empire and Portugal. In 1577, Elizabeth I sent an embassy led by Sir Edmund Hogan to Sultan Abd al-Malik I to open Moroccan markets to English merchants and explore the possibility of establishing a base on the Atlantic coast to threaten the Canary Islands and attack Spanish and Portuguese ships travelling to the Americas.[29] However, negotiations were interrupted a few months later by a dynastic crisis that culminated in the Battle of the Three Kings in 1578. In this battle, Abd al-Malik I faced his uncle, who was backed by King Sebastian I of Portugal. The battle resulted in the deaths of both claimants and the Portuguese king. The sultanate then passed to Abd al-Malik's brother, Ahmad al-Mansur. Another consequence of the Battle of the Three Kings was that Sebastian I died without an heir, plunging Portugal into a dynastic crisis that extended across the Strait of Gibraltar. One of the candidates for succession was António, Prior of Crato. Both England and the Netherlands sent ambassadors (along with costly gifts) to Sultan Ahmad al-Mansur to persuade him to support Crato over Philip II of Spain. In response, Philip II dispatched his own embassy (with equally costly gifts) to warn the young sultan about the imprudence of supporting Crato, saying how easily a member of the Sa'adi family could be found to challenge his throne. Ahmad al-Mansur, though young, understood Spain's implicit threats and spent subsequent years playing off both sides diplomatically. In 1585, Elizabeth managed to establish the Barbary Company in Morocco, granting England exclusive trade rights for 12 years. However, Ahmad al-Mansur never dared to directly challenge Philip II and refused logistical or financial aid when Elizabeth requested support for Drake's 1589 expedition to the Iberian Peninsula. The trade agreement was renewed in 1597, and plans for an invasion of Spain from Morocco were even considered. Ahmad al-Mansur dreamed of restoring Al-Andalus under his leadership,[30] but this plan never materialised beyond a personal aspiration of the sultan.

In summary, Elizabeth's rapprochement with the Ottoman Empire and Morocco was perceived by the Spanish government as both a provocation and a threat. Although neither Murad III nor Ahmad al-Mansur directly raised their scimitars against Philip II, this looming threat contributed to the Spanish king accelerating plans for the invasion of England in 1588.

António of Crato, a Crown, and the Portuguese Question

The Portuguese succession crisis was not a direct cause of conflict between England and Spain. Nevertheless, I have chosen to include it in this section for the following reasons: Elizabeth, much like Henry III of France, would use the pretender António, Prior of Crato, as a pawn in her strategic game against Philip II, reserving him when she felt politically vulnerable but employing him as leverage when she perceived strength, always seeking to extract the greatest possible economic advantage from the unsuspecting claimant. Moreover, this episode serves as a prelude to explaining several crucial developments in the subsequent war, and ultimately, to outlining the Spanish military doctrine that would later be reiterated in the 1588 Armada campaign. As previously noted, during the Battle of the Three Kings, King Sebastian I of Portugal died in combat. In August 1578, his great-uncle Henrique, then Archbishop of Lisbon, was crowned as his successor. The new king, nearly 70 years old and bound by ecclesiastical vows, had no recognised heirs. Thus began a succession contest featuring three grandsons of King Manuel I of Portugal: Catherine of Portugal, Duchess of Braganza, whose candidacy was weakened by her gender; Don António, Prior of Crato, the illegitimate son of Luís of Portugal, Duke of Beja;[31] and Philip II of Spain, whose mother, Isabella of Portugal, was a daughter of Manuel I. Under pressure from the Portuguese nobility, Catherine withdrew from the succession race. Don António, however, endeavoured to prove by all possible means that his parents had lawfully married,[32] thereby legitimising his claim to the throne. Despite his efforts, it was ultimately demonstrated that the evidence he submitted was fabricated and that the witnesses he presented had been bribed. Naturally, the Portuguese succession crisis was widely discussed in European courts, and England was no exception. Sir Francis Walsingham dispatched one of his most trusted envoys, Sir Edward Wotton, 1st Baron Wotton, to Lisbon, both to offer congratulations to King Henry upon his accession and to gain first-hand insight into the political situation on the Iberian Peninsula. Particularly noteworthy is a letter Wotton sent to Walsingham on 18 August 1579, in which he assessed the prospects of the three contenders to succeed the aging Henrique. Regarding António of Crato, he wrote the following:

> The things which are to hinder Don Antonio are the following. The King favours him not because of his dissolute life. He has many bastards by base women, most of them by 'new Christians'. It is feared therefore by the nobility that if he should come to be King, being unable by

ordinary means to make them all great he will seek to advance them by extraordinary means, and perhaps take dignities and 'incommiendas' from the rest of the nobility to give them. He is very poor, and therefore not able to win such of the nobility as are to be won by money; nor if it should come to force, would he be able to maintain a power in the field. Things which may further him are, that he is generally beloved of the people, gracious in his behaviour, and liberal in spending.[33]

In his letter, Wotton concludes that the candidate with the greatest likelihood of success is Philip of Spain. However, by the time Walsingham received the letter, events had already accelerated. On 31 January 1580, the cardinal-king Henrique passed away, leaving the Conselho de Governadores do Reino as regents. After deliberation, the council unanimously decided in mid-July that Philip would be the successor. Don António of Crato refused to accept this outcome and proclaimed himself king at the end of the month, amid popular acclaim. This development alarmed Philip, who was not prepared to let the opportunity to unite the Spanish and Portuguese crowns slip away. He therefore ordered the Duke of Alba to lead an army to seize the Portuguese throne by force. In response, António of Crato dispatched ambassadors to various European courts, requesting troops and weapons to counter the Spanish threat. On 19 August, Lord Leicester received a letter from António's ambassador, João Rodrigues de Sousa,[34] in which he requested the following:

> These are the things that the King of Portugal asks of the Queen of England.
> 1. Twelve ships well equipped with artillery, men and munitions.
> 2. Two thousand harquebusiers with their officers, who will all be paid in Portugal from the day of leaving England till their return.
> 3. As much bronze ordnance of all sorts as the Queen may be willing to direct; the realm being in great need of it.
> 4. A thousand quintals of gunpowder.
> 5. Two thousand quintals of iron balls of every sort.
> Payment will be made in Portugal, in coin, jewels, or specie as the Queen shall please.

In a letter from Walsingham to Cobham, dated 30 August, we learn that Elizabeth I held a secret meeting with João Rodrigues de Sousa.[35] However, he was unaware of what had been discussed or whether the Queen had committed herself to Don António's cause. What transpired was that Elizabeth, in a shrewd political manoeuvre, made vague promises, thereby gaining time to observe how the situation would evolve. Ultimately, Crato's army confronted that of the Duke of Alba near the banks of the Alcântara River, close to Lisbon, on 25 August 1580. The Portuguese forces, under the command of the Count of Vimioso, numbered 25,000 infantry and 2,500 cavalry. Opposing them, the Duke of Alba fielded 18,000 of his formidable tercios and 1,800 horsemen. The battle was brief. The Portuguese soldiers – mostly recently-conscripted peasants – stood little chance against what was considered the finest infantry in the world, which, upon crossing the river, massacred them.[36] Portuguese losses exceeded 4,000 men, compared to 500 on the Spanish side. Off the coast of Lisbon, António de Crato's fleet, commanded by Gaspar Brito, surrendered to the forces led by the Marquis of Santa Cruz. As the Duke of Alba's army entered Lisbon, António de Crato and the Count of Vimioso fled – first to the Azores archipelago, where five of its seven islands pledged allegiance to him, and later to France and England. There, he hoped that the monarchs, hostile to Philip of Spain, would support him militarily in reclaiming the throne. Relations between Don António and England intensified from April 1581, when preparations began for an expedition to the Azores. This moment proved pivotal, not only for the commercial (and imperial) ambitions of the various European powers. The Azores were the key to everything – the key to the treasure. This is best understood by considering the prevailing wind patterns in the Atlantic. In the Canary Islands region, winds blow from east to west, while further north, at the latitude of the Azores, they reverse direction – from west to east. In the age of sail, these wind systems were inescapable, and, together with Atlantic currents, made transatlantic voyages possible. However, the journey was so long that it necessitated stopovers at strategic points to take on fresh water and food (typically fruit and fish). For instance, the Spanish treasure fleets, on their outward journey from Spain, would first stop in the Canary Islands and, on their return, in the Azores. Similarly, English privateers followed the same route.

If the Azores were to fall under the control of a power hostile to Spain, it would have signified, quite simply, the end of its empire. Meanwhile, António of Crato continued to grant concessions to the English Crown,[37] from which he received little more than vague promises and minimal support – such as two ships and a garrison of approximately 100 men under the command of Captain Henry Richards and John Sachfield, tasked with the defence of Terceira.[38] At the French court, Don António found greater success. Although King Henry III was not prepared to declare war on Spain,[39] António's ambassador in Paris, the Count of Vimioso, benefited from the support of Catherine de' Medici, the Queen Mother. She agreed to back a military campaign against the Azorean islands loyal to Philip II in exchange for significant commercial privileges in the Americas and the transfer of control over present-day Brazil. However, it was Spain that made the first move toward recovering the rebellious islands. In June 1581, orders were issued for Pedro López de Valdés to organise a fleet to conquer Terceira, the most strategically important of the Azores. Yet the admiral disobeyed the King's instructions, and his attempt to seize the island failed. This attack alerted Spain's rivals to the seriousness of the situation. The matter began to be resolved in July 1582, when the time came to determine the true master of the Azorean archipelago. General Commander Filippo Strozzi[40] had spent several months preparing a fleet for the campaign. However, to Don António's dismay, Elizabeth I failed to honour her commitments and refused to participate in the expedition,[41] even though Francis Drake had expressed interest in joining. Pressured by Bernardino de Mendoza, Spain's relentless ambassador, Elizabeth limited her support to the shipment of cannons and ammunition and granted Don António permission to recruit English privateers willing to join Strozzi's fleet. Ultimately, he succeeded in hiring seven ships and their respective crews:[42] *Diamond, Archangel, Prosperity, Greyhound, White Bear, Antonia* and a seventh vessel whose name remains unknown. These were commanded by privateers such as Thomas Walton, Clinton Atkinson, Thomas Beavyn, and John Storye, who were joined by Richards and Sachfield with their two ships already stationed in Terceira. The two fleets finally encountered one another in the Azores on 22 July 1582, engaging in four days of combat in what became known as the Battle of Vila Franca do Campo, also referred to as the Battle of Ponta Delgada or even the Battle of São Miguel. Don António's fleet, comprising approximately sixty vessels, was under the command of Filippo Strozzi. It also included the pretender himself, the Count of Vimioso, and several prominent French nobles such as the Count of Brissac and General Sainte-Souline, who commanded the infantry forces. The Spanish fleet was under the command of the Marquis of Santa Cruz and included the presence of Admirals Cristóbal de Eraso – his second-in-command and Captain General of the Indies Fleet – and Miguel de Oquendo, while the marine infantry was led by Field Master Lope de Figueroa. Traditionally, studies of this battle have focused on the events of 26 July, which marked its climax. It is not the purpose of this book to recount the battle in exhaustive detail;[43] however, there are two episodes from these days that merit closer examination for the reader to better understand Spanish military doctrine, and how it would later be applied by the Armada of Medina Sidonia in 1588. The first episode occurred on 24 July. That day, the wind favoured the French, who took the windward position. They formed into three squadrons (see the corresponding plate Elementary Tactical Schemes I) and launched an attack on the Spanish rearguard, commanded by Miguel de Oquendo, whose ships were moving dangerously close to the coast of the island of Santa Maria, where they risked becoming trapped with little room for manoeuvre. At that critical moment, the Marquis of Santa Cruz ordered his ships to luff (turn their bows into the wind to present their broadsides to the French), thereby positioning themselves to fire their artillery. In other words, the Spanish were about to make tactical use of long-range artillery to prevent Strozzi's ships from approaching. The French admiral, adhering to the chivalrous conventions of the time, chose to tack and attack the Spanish vanguard – his flagship was expected to engage the Spanish flagship, namely Santa Cruz's *San Martín*. In response, the Spanish commander ordered his ships to turn, resulting in the Spanish vanguard sailing from south to north, while the French fleet moved from north to south. Upon witnessing this manoeuvre, Oquendo also turned to rejoin the vanguard, and the following occurred, as described in the Basque admiral's report:

> And when the Marquis saw the enemy's audacity, he placed himself athwart with the King's two galleons, *San Martín* and *San Mateo*, and I aligned mine in formation, and we took the rest of our fleet under our protection. Thus, arranged in this sound order, the enemy passed windward of us with all of his heavy ships, firing the entire broadside artillery. And as the galleons, being well-armed, returned fire, there was a good skirmish – though there was no musket or arquebus fire – and thus the day came to an end.

In other words, what we witness here is a line-of-battle artillery engagement executed by the Spanish fleet. This episode has largely gone unnoticed, principally because it is not mentioned in the French report of the battle – the version that ultimately reached England. However, modern historians could easily find references to it in the two reports composed by the Marquis of Santa Cruz and Miguel de Oquendo. That they have not done so may be attributed to ignorance, neglect, disdain, or simply scholarly inertia.

The second episode of this battle that warrants close attention took place on its decisive day, 26 July. The protagonist would be the galleon *San Mateo*, the very same vessel that would later feature in the Armada of 1588. On that morning, Strozzi's fleet once again secured the windward position and launched an assault on the Spanish fleet, whose vanguard consisted of its four most formidable vessels: the hulk *San Pedro*, commanded by Bobadilla; the galleon *San Martín*, under Bazán himself; the *Jesús María*, commanded by Eraso; and the galleon *San Mateo*, led by Lope de Figueroa. Further behind was the rearguard, under the command of Miguel de Oquendo. Thus, the vanguard was composed of four heavily armed ships that were difficult to board, especially considering that the French ships, like their English counterparts, were of lighter tonnage. At a certain point, the *San Mateo* became separated from the Spanish formation. The reasons for her isolation remain unclear: it is uncertain whether this was a coordinated tactical manoeuvre agreed upon with Bazán, a unilateral initiative taken by Lope de Figueroa, or simply the result of some unexpected mishap. Regardless, Strozzi and his fleet seized the opportunity and attacked the *San Mateo*, in a fashion reminiscent of numerous similar situations that the Armada of 1588 would face in the English Channel. It is worth noting that if one ship could effectively distract the French fleet, it was one of the two Spanish galleons. Bazán's flagship, however, could not be risked in such a manoeuvre, as he was responsible for command and coordination. Thus, the task fell to Figueroa's *San Mateo*, which carried the finest and most seasoned soldiers of the marine infantry. Coincidence? Suddenly, the *San Mateo* found herself surrounded by five French ships, including those of Strozzi and the Count of Brissac. Meanwhile, Bazán's squadron continued to advance. Pursued by the French right wing, it tacked and, once again positioning broadside, commenced a fierce artillery barrage, sinking another French vessel in the process. One of the critical turning points of the battle occurred when, with both Strozzi's flagship and his vice-flagship stubbornly engaged in attempting to board the *San Mateo*, the French fleet was left without effective command during the heat of combat. This state of affairs persisted for over two hours, during which the five French vessels failed to capture the *San Mateo*. At the same time, Bazán's squadron inflicted heavy damage on the enemy's right wing through sustained cannon fire. From the rear, Oquendo's squadron had sailed to leeward and managed to position itself behind the French rearguard, seizing the windward advantage. To compound the French predicament, Sainte-Souline, commander of the left wing, had vanished from the battlefield – deserting with his entire contingent. Now, with the windward position secured, Oquendo launched his entire force into the melee surrounding the *San Mateo*, a manoeuvre also attempted by the *San Martín*, albeit at a much slower pace due to adverse winds. Thus, it was the *Juana*, from Oquendo's squadron, that reached the fray first, grappling the stern of Strozzi's *Saint Pierre*, while the *Jacqueline*, commanded by the Count of Brissac, was caught by Oquendo's *Catalina*. The rest is history. Though the *Saint Pierre* ultimately received reinforcements and managed to disengage from the *Juana*, her attempt to retreat was thwarted by the arrival of Bazán's *San Martín*. The battle could now be considered concluded. Bazán had secured yet another epic victory to add to his distinguished record – against a fleet that outnumbered his in sail and under unfavourable winds. The French had lost ten ships and approximately 2,000 men, among them the notable figures of Strozzi himself and the Count of Vimioso, António de Crato's chief lieutenant. In the days that followed, at least three additional vessels were abandoned on the island of Terceira, and there is documented evidence that ocean currents carried another ship to the coast of Portugal, its entire crew found dead on board.[44] The claimant, who had remained in refuge on Terceira during the battle, attempted in October to launch an assault on Madeira, relying on two English ships[45] – presumably those of Richards and Sachfield. However, according to the French historian, citing his colleague De Thou,[46] these ships were, quite literally, lost, leading the claimant to abandon his plan to attack Madeira and return to France. It is also worth noting the involvement – or rather, the limited involvement – of the English privateers in this campaign. In fact, 'involvement' may be too strong a word, as it appears they either refrained entirely from entering combat or confined themselves to long-range cannon fire with Bazán's fleet. They did, however, manage to return safely to England. Nonetheless, several of them would not live much longer. This was the case for Walton and Atkinson, who often operated as a pair. In the summer of 1583, they had attacked the Scottish vessel *Grace of God* and subjected its crew to brutal torture, an act that provoked formal complaints from King James VI of Scotland. Queen Elizabeth subsequently ordered their arrest, and both were tried, condemned, and executed by hanging on 30 August of that same year.[47] Beavyn would share a similar fate some years later. The conquest of the Azores was completed on 2 August 1583 in an amphibious operation described by John F. Guilmartin as 'strictly modern'.[48] The campaign was once again led by the Marquis of Santa Cruz, who even designed the landing craft used in the assault on Terceira. With only minor modifications – and constructed entirely of wood – these vessels bore an extraordinary resemblance to those employed by the Allied forces during the Normandy landings in 1944. It was a remarkable operation, executed with overwhelming success, and closely followed from London.[49] For several months, António de Crato and his agents continued to send letters to England claiming that the Spanish had failed in their attempt to seize the island. It was not until 19 October of that year that he formally acknowledged defeat.[50] The claimant continued to reside in France for a time, under the protection of Catherine de' Medici. However, his position in France eventually became untenable, prompting him to take the first steps toward establishing himself in England. Accordingly, he entrusted the money and jewels he still held in France to Sir Francis Drake for safekeeping.[51] On 30 January 1584, he sent a formal letter to Queen Elizabeth requesting permission to reside in her kingdom.[52] The Queen granted his request, though she refused to accede to certain additional petitions,[53] and the claimant ultimately relocated to England.

In summary, throughout this chapter we have observed that António de Crato was not among the principal factors leading to the outbreak of war, although he would later have his moment of significance. For Queen Elizabeth, the Portuguese claimant was merely a pawn in the wider chess game being played with Philip II. Until 1585, the Queen did not dare confront Spain openly, and during this period, the veiled threats of Bernardino de Mendoza weighed more heavily on her decisions than the concessions offered by António. What is undeniable is that Elizabeth was an exceptionally astute ruler; to her, the Portuguese claimant was a card to be played – an asset to be deployed, and if necessary, sacrificed, at the opportune moment. In the meantime, however, she could exploit him economically – renting ships, selling him artillery and munitions – leaving Catherine de' Medici to assume the burden of direct confrontation with Spain. In this chapter, we have also begun to examine the operational principles of Spanish naval doctrine – an aspect that will be revisited and explored in greater depth when we turn to the Armada of 1588. Most notably, we have analysed the conquest of the Azores. This archipelago held the key to transatlantic commerce between Europe and the Americas. The two remarkable campaigns led by Álvaro de Bazán enabled Philip II to consolidate his empire. These campaigns, ultimately, ensured that Spain would retain its imperial presence in the Americas for at least another two and a half centuries.

Mary Stuart, complots and the Catholic Question

One of the defining factors of Elizabeth's policy was religion – more specifically, the consolidation of Anglicanism, the repression of Catholicism, and her continuous tensions with the papacy. It is, indeed, paradigmatic that while Queen Mary has gone down in history with

the moniker 'Bloody Mary' due to her persecution of the Anglican faith –
during which approximately 280 Protestants were burned at the stake –
Elizabeth's treatment of Catholics was not fundamentally different
from that of her predecessor, though she has never been labelled
'Bloody Betsy'. At the outset of her reign, Elizabeth reinstated the laws
that established Anglicanism as the sole official religion. Nevertheless,
it is unlikely that religious motivations were the primary factor
inducing Philip II to contemplate an invasion of England. However, it
is plausible that he felt considerable pressure from successive popes
to act against Elizabeth. It is worth recalling that Henry VIII's 'divorce'
from Rome did not prevent the devoutly Catholic Charles I of Spain
from maintaining excellent relations with the notoriously uxoricidal
English monarch, not even after the Pope excommunicated him in
1533. In contrast, we may affirm that Philip II astutely exploited Rome's
obsession with Elizabeth to persuade the papacy to provide financial
support for his 'Enterprise of England'.

Tensions between the Queen and the papacy escalated from the
very moment of Elizabeth's accession. Determined to reinvigorate the
religion founded by her father, she swiftly enacted a series of laws –
including the famous Acts of Supremacy and the Recusancy Acts of
1559 – that imposed fines, imprisonment, and other penalties on those
who failed to attend Anglican services. Unsurprisingly, discontent
spread among the still sizeable Catholic population, which found itself
forced to practise its faith in secrecy and under increasing surveillance.
In May 1568, a pivotal event occurred: Mary Stuart, having escaped
from imprisonment in Scotland after being compelled to abdicate,
arrived in England. She hoped that her cousin Elizabeth would honour
her promise to help restore her to the Scottish throne or, at the very
least, grant her protection. Mary's arrival placed Elizabeth in an
extraordinarily delicate position, as Mary could be seen as the most
legitimate successor to the English crown, especially given Elizabeth's
lack of an heir. The Queen feared that her cousin might seek to hasten
her own accession through conspiracy or intrigue. These anxieties
were exacerbated by Mary's Catholic faith and the possibility that her
political programme might involve the restoration of papal authority
in England. As a result, Mary Stuart was placed under confinement –
initially in Carlisle Castle in Cumbria, and later in Bolton Castle in
Yorkshire. At this juncture, a third woman enters the stage.

This woman was Jane Howard, Countess of Westmorland. She
devised a plan to marry her brother, Thomas Howard, Duke of Norfolk,
to Mary Stuart and thus initiate a political alliance aimed at deposing
Elizabeth. To that end, she orchestrated the uprising in November

1569, in which her husband, Charles Neville, Earl of Westmorland, along with Thomas Percy, Earl of Northumberland, took up arms and marched south in what became known as the Rising of the North, or the Revolt of the Northern Earls. However, the rebellion proved a complete failure – not only due to its poor organisation, but also because it failed to garner any popular support. Indeed, this would become a recurring theme in subsequent conspiracies and even in Spanish invasion plans: the mistaken belief that English Catholics would rise up against their Queen. Yet such a Catholic uprising never materialised. It is significant that, despite the repression of Catholicism, adherents of the persecuted faith remained loyal to Elizabeth over the Pope – even English Catholics living abroad. A paradigmatic example is that of Sir Richard Shelley, who in 1559 was appointed Grand Prior of the Knights of Saint John in England while on a diplomatic mission in Italy. However, he never assumed the post at the Queen's request. In fact, by 1569, he had settled in Venice, where he engaged in intelligence work on behalf of Elizabeth.[54] Returning to the fate of the rebellious earls: they fled to Scotland. While Westmorland was betrayed and handed over to the English Crown – subsequently being executed by order of the Queen – Northumberland managed to escape to Spanish Flanders, from where he hoped to organise another rebellion. He died in exile in 1601. As for the Duke of Norfolk, whose involvement in the uprising was ambiguous, he spent several months imprisoned in the Tower of London. From this moment onward, events accelerated. It became clear that Mary Stuart posed a genuine threat to Elizabeth, despite appearing uninvolved in the Countess of Westmorland's machinations. Moreover, Pope Pius V issued the bull *Regnans in Excelsis*, excommunicating Elizabeth and authorising Catholics to disobey her. This papal edict triggered a renewed wave of repression against Catholics – particularly Jesuits – but also emboldened a faction of English Catholics to conspire against the Queen in what became known as the Ridolfi Plot. This conspiracy was conceived by the Florentine banker Roberto Ridolfi and comprised three coordinated actions: an invasion of England by the Spanish viceroy in Flanders, the Duke of Alba, commanding 10,000 troops; the instigation of a Catholic uprising within England; and the marriage of Thomas Howard, Duke of Norfolk, to Mary Stuart. The exact means by which the plot was uncovered remain uncertain. Some believe that John Hawkins, the English naval commander, led the Spanish ambassador Guerau de Spes and Ridolfi to believe he was willing to participate, thereby extracting their confessions. Others suggest that Ridolfi himself may have been a double agent, working to discredit Catholics and potential adversaries of the Queen such as

Norfolk, who was ultimately imprisoned and executed, and de Spes, who was expelled from England – an interpretation supported by Ridolfi's biographer, L. E. Hunt.[55] Spanish archives, however, contain two documents that raise suspicions regarding Philip II's intentions toward England – not necessarily to depose Elizabeth, but rather to pressure her into restoring Mary Stuart to the Scottish throne.[56]

What is irrefutable is that this plot did not improve Anglo-Spanish relations, nor did it advance the cause of Mary Stuart. In 1580, the Jesuits pressured Pope Gregory XIII to de-escalate tensions in England, prompting the Holy Father to declare that English Catholics should not disobey the Queen in civil matters, though they were still bound to resist in matters of faith. In 1583, yet another conspiracy was uncovered – the so-called Throckmorton Plot, orchestrated by Francis Throckmorton. His family had held a prominent place at the English court since the reign of Henry VIII, being related to his sixth wife, Catherine Parr. His uncle, Nicholas Throckmorton,[57] had served as a diplomat under Elizabeth, and his cousin Bess Throckmorton was a Gentlewoman of the Privy Chamber to Queen Elizabeth and later the wife of Walter Raleigh. This plot bore many similarities to the Ridolfi conspiracy, although in this instance, the role formerly intended for Norfolk was to be played by the Duke of Guise, who would invade England from France with a Franco-Spanish army and marry Mary Stuart. The plan was uncovered by an English spy operating within the French embassy in London, who informed Elizabeth's principal secretary, Sir Francis Walsingham. Once again, arrests and trials ensued, though only Francis Throckmorton was executed. In early 1585, the so-called Barry's Plot unfolded, in which William Barry, in collaboration with Edmund Neville, planned to assassinate the Queen due to her anti-Catholic policies. However, Neville eventually betrayed the plan, leading to Barry's arrest, trial, and execution. As previously noted, 1585 marked the beginning of the war between England and Spain. It was, however, a war never formally declared – rather, it was a gradual escalation of hostilities between the two powers. It was in this year that Philip II ordered Álvaro de Bazán and Alessandro Farnese to begin devising the invasion plan that would culminate in 1588. Meanwhile, the repression of English Catholics intensified. That same year, Parliament passed the Act Against Jesuits and Seminarists, which led to the execution of numerous priests, deacons, and lay adherents of the Catholic faith. It is estimated that more than 189 individuals were executed for their Catholic beliefs during Elizabeth's reign, of whom 20 have since been canonised. The defeat of the Spanish Armada in 1588 was followed by a wave of arrests and executions. This section concludes with the Babington Plot of 1586 – a renewed attempt, led by Anthony Babington, to assassinate Elizabeth and place Mary Stuart on the throne. It appears, however, that Walsingham, Elizabeth's spymaster, uncovered the conspiracy in its earliest stages and allowed it to develop to entrap Mary Stuart, who was then incommunicado at Tutbury Castle. The goal was to secure irrefutable evidence of her complicity and confirm her as a threat to the English Crown. Babington and thirteen other conspirators were arrested, tried and executed. Mary Stuart's situation became untenable, and she too was tried for treason and sentenced to death. To the astonishment of all Europe, she was executed by beheading on 8 February 1587. Some historians argue that Mary Stuart's execution prompted the invasion plans against England, but as noted earlier, preparations had already begun well before. In any case, in the deeply religious world of the sixteenth century, the conflict between Catholics and Anglicans remained a constant source of political tension between England and Spain – a tension that would persist throughout the seventeenth century.

Another crucial front that must be addressed is the French Wars of Religion, which took place between 1562 and 1598 and are commonly divided into eight distinct military conflicts. Despite Catherine de Medici's determined efforts to achieve a negotiated peace, the opposing factions – led by the Bourbon and Guise families – refused to yield, particularly during the reigns of Francis II and Charles IX. Both Elizabeth I and Philip II felt compelled to defend their respective political and religious positions, and from the very onset of the conflict, they supported their allies financially and militarily, even going so far as to send military contingents into French territory. On 10 June 1584, Francis, Duke of Anjou – former suitor of the resolutely unmarried Elizabeth I, but more significantly, the last remaining heir to his brother, King Henry III – passed away. Given that Henry III had no offspring and was widely suspected of lacking the intention or capacity to produce any, his brother's death precipitated a succession crisis. The Catholic League seized this opportunity to negotiate the Treaty of Joinville with Philip II of Spain on 31 December 1584. Under the terms of the treaty, Spain committed to supporting the Catholic cause in France, in exchange for several concessions: the severing of France's alliance with the Ottoman Empire, the cessation of French incursions into the Azores, and the restoration to Spain of territories in the Low Countries that the United Provinces had previously ceded to France. Furthermore, the treaty accepted Charles de Bourbon as heir to the French throne, thereby excluding the Protestant Henry of Navarre. However, the true turning point in this period came with the

Treaty of Nemours, signed on 7 July 1585. This agreement effectively subordinated the interests of the French crown to those of the Catholic League, led by the Duke of Guise, thereby formalising and reinforcing the commitments previously established between Guise and Philip II. Although this treaty would soon spark the War of the Three Henrys, it also raised significant concerns in London. The prospect of a Catholic victory in France, a country increasingly seen as a potential ally of Spain, alarmed Elizabeth I. Consequently, she not only continued to support the Protestant faction in France but also recognised the urgent need to intervene more decisively in the other major theatre of Spanish conflict: the Netherlands.

William of Orange, Nonsuch and the Dutch Question

Given the complexity of the situation in the Low Countries, understanding why Philip of Spain was present there and why he found himself at war requires a retrospective look into earlier centuries. Specifically, it is necessary to go back to the fourteenth and fifteenth centuries, when the Dukes of Burgundy, through a combination of conquests and strategic marriages, expanded their dominion into the region roughly corresponding to what is now known as the Benelux. In 1477, following the death of Duke Charles, known as the Bold, the title passed to his only daughter, Mary the Rich. That same year, the King of France, Louis XI, militarily occupied the Duchy of Burgundy. As a result, paradoxically, the Burgundians lost the territory that had given them their ducal title but retained control over the Low Countries, also known at the time as the Seventeen Provinces. Mary married the powerful Maximilian of Habsburg, and from this union was born Philip the Handsome, future husband of Joanna of Castile (daughter of the Catholic Monarchs). Their son would become one of the most pivotal figures in European history: Emperor Charles V. His inheritance was monumental: the Kingdoms of Aragon, Castile, Naples, Sicily, Sardinia, Navarre, Austria, and Granada, the territories of Artois, Luxembourg, the Seventeen Provinces, and the Franche-Comté, along with the rights to the Holy Roman Empire and the Duchy of Burgundy. In addition, he held various possessions in North Africa . . . and the New World. After a long reign, between 1555 and 1556, Charles decided to abdicate and retire to a monastery in the Spanish region of Extremadura. He divided his vast empire between his brother, Ferdinand, and his son, Philip. The former received Austria and the title of Holy Roman Emperor, while the latter inherited all the remaining territories, including the Seventeen Provinces. It is essential to recognise that this was not a homogeneous territory. It comprised predominantly agrarian regions as well as cities of exceptional commercial dynamism, such as Antwerp, Ghent, Bruges, and Brussels – urban centres where Catholics and Protestants coexisted. Furthermore, the region held great strategic significance: it bordered France, Spain's main rival at the time, and included the mouths of the Rhine and Meuse rivers – key arteries for European trade. Philip II's governance over these territories was fraught with difficulties from the outset. His religious intransigence (manifested in the severe repression of Protestant worship), the heavy fiscal burdens imposed on the population, and the continuous presence of Spanish troops (given that Spain remained at war with France) all provoked discontent among the Flemish princes. They increasingly perceived Philip II as a foreign monarch who posed a threat to their autonomy.

The governor of the territory was Margaret of Parma,[58] an intelligent and capable woman who, moreover, had been born in Flanders and was thoroughly familiar with the region and its particularities. However, she lacked the necessary means to temper the discontent of the rebellious nobles, who ultimately rose in arms in what would become known as the Eighty Years War – a true Vietnam for the Spanish Crown. Successive governors[59] alternated between a policy of iron-fisted repression and diplomatic overtures, yet they were never able to subdue the obstinate Protestant princes who refused any form of truce or surrender. In 1576, Philip II appointed his half-brother, John of Austria, as the new governor of the Low Countries. However, upon his arrival, he encountered a chaotic situation: the States General – the governing body dominated by the Protestant north – refused to recognise his authority, forcing him to negotiate a peace settlement. This agreement, signed on 7 January 1577, came to be known as the Perpetual Edict. Under its terms, the States General agreed to recognise Philip II as king and John as governor, in exchange for the withdrawal of all foreign troops and the granting of a general amnesty, among other concessions. However, the Edict posed a new challenge for the newly appointed governor, as without a military force none of his directives were obeyed. Moreover, he uncovered a plot to assassinate him, compelling him to abandon Brussels and seek refuge in Mechelen. In early December, the States General declared John 'an enemy of the fatherland'. Besieged at the Castle of Namur, he appealed to Philip II for reinforcements. In response, the king dispatched 20,000 soldiers under the command of his nephew and one of the central figures in the history of the Spanish Armada: Alessandro Farnese, Duke of Parma. This brought together two brilliant individuals who, in addition to their shared military prowess, had known and cherished each other since youth. They had grown up together at the Spanish court and later received an elite education at the University of Alcalá de Henares. They had also fought side by side at the Battle of Lepanto, where glory was chiefly attributed to John of Austria (notwithstanding the pivotal role of Álvaro de Bazán, the Marquis of Santa Cruz).

We must now cross the English Channel to briefly visit the court of Queen Elizabeth. Until that point, the Queen and her advisers had followed the developments in the Low Countries with great attention. Indeed, the rebel cause led by William of Orange elicited considerable sympathy in London. From a religious perspective, the rebels were opposing Catholic forces; from a geopolitical standpoint, however, a Dutch defeat would place Elizabeth's foreign policy in a precarious position. As in the case of António of Crato, the Queen's council was divided. One faction, which might be referred to as the 'War Party', led by Walsingham and Leicester, advocated for the deployment of English troops in support of the rebels, even at the risk of open war with Spain. In contrast stood those favouring official neutrality, headed by Lord Burghley, a stance Elizabeth chose to adopt during the early years of the conflict. Nonetheless, she granted permission for as many volunteers as wished to fight in Flanders, provided it incurred no expense to the Crown. Thus, the first English expedition set out in April 1572, under the command of Thomas Morgan, comprising a modest force of 300 men. A few months later, Morgan persuaded Sir Humphrey Gilbert to join with additional volunteers, raising the total number of troops by another 1,500. The year 1572 is widely regarded as a true *annus horribilis* for the Spanish governor of the region, the Duke of Alba, who received scant resources to counter the Dutch army and its allies. However, financial difficulties also plagued the rebels, who in addition had to manage tensions between their English and French supporters. These financial constraints led to the desertion of many foreign soldiers, including Gilbert. Yet there was never a shortage of men willing to take their place – such as the regiment organised by John Norris[60] in 1577, which included veterans of earlier expeditions like the bellicose Roger Williams. It is also worth noting the significant presence of Scottish soldiers,[61] particularly following the end of the Marian Civil War (1568–73).

At the beginning of 1578, a turning point in the war occurred when Spanish and Dutch forces clashed near Namur, south of Brussels, at the Battle of Gembloux. Approximately 25,000 Dutch troops under the command of Antoine de Goignies faced 17,000 Spanish soldiers – many of them seasoned tercios – led by John of Austria, with Alessandro Farnese commanding the cavalry. The Spanish achieved a crushing victory, inflicting over 10,000 casualties on the rebel army, including dead, wounded, and captured, while suffering only 20 losses themselves. The triumph proved decisive, reversing the course of the war (and arguably prolonging it for another 60 years), and triggered a wave of Spanish conquests:[62] Hainaut, Namur, Brabant and Luxembourg returned to royal obedience. The victory at Gembloux fractured the unity among the rebel provinces and alarmed the court in London, which responded by hiring the mercenary army of Duke Casimir, son of the Elector Palatine, to intervene in the Low Countries, launching an attack on Spanish forces from the east. Simultaneously, the Duke of Anjou opened a new front from the south, advancing from France. To further complicate the situation, on 1 October 1578, John of Austria died of dysentery. He was succeeded by Alessandro Farnese, Duke of Parma, who would soon prove himself not only an exceptional military commander but also a skilled diplomat. Farnese defeated both Anjou and Casimir and succeeded in securing control over the southern provinces, which, in January 1579, formed the Union of Arras, pledging allegiance to Philip II. In response, a few weeks later, the northern provinces founded the Union of Utrecht, vowing disobedience to the Spanish Crown. The creation of these two alliances marked the de facto division between the Catholic south and the Protestant (and nominally independent) north. Meanwhile, John Norris remained highly active in the Dutch campaigns. He was present at Gembloux and collaborated in various operations with the Duke of Anjou. He achieved notable victories, such as at the Battle of Rijmenam (1578) and the conquest of Mechelen (1580), but also suffered significant defeats, including at Borgerhout (1579) and Noordhorn (1581). Nevertheless, Norris gained fame and recognition at Elizabeth's court, where he was regarded as a charismatic and capable commander. The period from 1579 to 1585 marked the steady collapse of William of Orange's armies, with particularly significant victories by Parma in Lier (1582), Eindhoven (1583), Steenbergen (1583), Ghent (1584), and Zutphen (1584). In July 1584, the siege of Antwerp began – a strategically vital city whose fate could determine the outcome of the war. In 1580, the United Provinces signed the Treaty of Plessis-lès-Tours with France, naming the Duke of Anjou as protector of the rebel territories. However, he was unable to withstand Parma's military pressure and returned to France in 1583. On 10 July 1584, William of Orange was assassinated by Balthasar Gérard, a fanatical Catholic. Notably, the English officer Roger Williams was present at the scene of the assassination; although he was unable to prevent the stabbing, he played a role in apprehending the murderer.[63] By 1585, the situation had become so dire that the States General offered sovereignty over their territories to Henry III of France in exchange for intervention in the conflict. On 9 March, the French king declined the offer. In May of that year, Elizabeth I received the same offer, which she too refused; however, she did not hesitate to increase her support for the Dutch rebel cause.[64] Determined to take the final step toward war with Spain, she concluded the Treaty of Nonsuch with Justinus of Nassau in August 1585, by which England officially pledged military support to the United Provinces. This official backing of military intervention in a territory under the sovereignty of the King of Spain constituted, in diplomatic terms, a declaration of war.

Chapter 3

1585–1588: EARLY BATTLES

As there was no formal declaration of war, historiography has generally agreed to mark the signing of the Treaty of Nonsuch as the official beginning of the conflict. As we saw in the previous chapter, over the years Elizabeth had pursued a policy of gradually increasing aggression against Spanish interests – justified, in her view, as a response to the attacks on English commercial enterprises, particularly the Spanish monopoly over trade with the New World, as well as blockades and embargoes. Added to this was a growing sense of threat from the numerous plots orchestrated by English Catholics against her life, for which the English court saw Philip II as the principal instigator and patron. Between 1585 and 1588, the first military engagements took place. These initial encounters reveal that England consistently seized the initiative, while Spain began preparations for a large-scale invasion. The following section analyses the key military actions of this period.

Drake in the Caribbean

In the spring of 1585, England began organising the largest privateering expedition yet undertaken in the Caribbean. The fleet consisted of twenty-one ships and eight pinnaces, carrying approximately 2,300 men. It was a privately funded venture, with an investment of £60,000 and an expected return of £600,000. Among the investors was Queen Elizabeth I herself, who contributed the *Elizabeth Bonaventure*, which served as the flagship of Francis Drake, and the *Aide*, captained by Edward Winter, son of William Winter, who also provided the *Sea Dragon*, under the command of Henry White. Drake contributed three ships of his own: the *Thomas*, captained by Thomas Drake; the *Elizabeth Drake*, under John Vaughan; and the *Francis*, commanded by Thomas Moone. The Earl of Leicester contributed two ships: the *Leicester*, a

400-ton vessel captained by his brother-in-law, Vice-Admiral Francis Knolles, and the *Speedwell*. Notably, among the vessels that participated in the expedition was also the *Primrose*, under the command of Martin Frobisher. The infantry was under the command of Lieutenant General Christopher Carleill,[1] who sailed aboard the *Tiger*, with Anthony Powell serving as sergeant major and colonels Matthew Margan and John Simpson among his principal officers. While the overarching goal of the expedition was to plunder various Caribbean settlements, the primary target was the city of Nombre de Dios. Drake, drawing on prior experience, knew that this city often served as the principal storage site for large quantities of gold and silver from Peru, awaiting shipment to Spain.[2] This, of course, was without discounting the opportunity to intercept and seize any isolated galleons or merchant vessels carrying valuable goods. Various logistical issues – which would persist throughout the campaign – delayed departure until 14 September. A week later, the fleet arrived off the coast of Galicia, anchoring at the Cíes Islands. The purpose of this brief stop was to resupply with fresh water and to inquire about the fate of several English ships that had been detained during the embargo of the previous May. In the port of Baiona, Drake was informed that most of the merchants had already returned to England, while those still anchored in Vigo were given the option of liberation. However, the merchants reported they were being treated well and preferred to remain to negotiate their own return independently.[3]

A week later, with barrels refilled with fresh water, Drake set sail once more. It is worth noting, however, that neither Baiona nor Vigo were attacked. Only a small chapel was looted, along with several boats belonging to residents who had attempted to flee upon the arrival of the English fleet. Nevertheless, thanks to a truce negotiated with the governor of Baiona, Pedro Bermúdez, all stolen goods were returned in exchange for fresh water. As previously stated, Drake resumed his journey after a week, hoping to intercept a galleon from the Spanish treasure fleet that might have fallen behind on its journey from the Azores to Seville. However, that year the voyage had proceeded more swiftly than usual, and no such vessel was encountered. The next destination was the Canary Islands, where Drake decided to launch an assault on Santa Cruz de La Palma, capital of the island of La Palma – one of the westernmost islands in the archipelago. Unbeknownst to him, Santa Cruz was the most heavily fortified port in the region, under the command of the capable and decisive Jerónimo de Salazar. The artillery from its three forts – Santa Cruz del Barrio del Cabo, San Miguel, and the principal fortress, Santa Catalina – together with the determined actions of the local militia, successfully repelled the English attack. The assault damaged the *Bonaventure* and *Leicester*, and inflicted the first casualties among Drake's infantry. Faced with this unexpected resistance, the expedition headed south to the uninhabited island of El Hierro to resupply. The following stop was the Cape Verde Islands – a curious decision that has intrigued many historians, as it signified a significant detour from the expedition's original objectives. Some have speculated that Drake may have been motivated by a desire for revenge after William Hawkins's failed expedition to the archipelago in 1583. Regardless, the English privateer showed no mercy. He sacked the towns of Ribeira Grande and Praia, seizing cannons, church bells and every valuable object he could find over a 14-day period, along with all food and water supplies that could be transported. However, the fleet also unknowingly acquired a deadly and silent enemy: malaria. The disease was carried by twenty enslaved Africans discovered in a hospital and brought aboard the English ships. By the time the expedition reached the Caribbean, approximately 300 crew members had died from the illness, and many more were ill. The *Bonaventure* and *Primrose* suffered the most losses, accounting for over half the fatalities. The fleet briefly rested on the uninhabited island of San Cristóbal (modern-day St. Kitts) before sailing toward Hispaniola to attack its capital, Santo Domingo – the oldest and largest city in the Caribbean, though by then already in decline. Fortune favoured Drake when he captured a ship whose pilot, a Greek sailor, provided detailed intelligence on how to attack the city, including the weakest points in its defences and strategies for avoiding the formidable cannons of the Ozama Fortress. On 11 January 1586 (New Year's Day according to the English calendar), Drake successfully captured the city in a matter of hours. Only the Ozama Fortress resisted briefly. The English remained in Santo Domingo for a month, during which they looted valuables worth an estimated 32,000 pesos and negotiated a ransom with Governor Ovalle, who had fled inland with the civilian population. Initially, the English demanded the enormous sum of 2,000,000 pesos, but Ovalle managed to raise only 50,000, which Drake ultimately accepted. Upon departing, the English destroyed churches, convents, and – most tragically for historians – the oldest Spanish archive in the Americas. The next target was Cartagena de Indias, a key city where much of the treasure intended for Spain was stored. However, by the time of Drake's arrival, the treasure ships had already departed. Governor Pedro Fernández de Busto fled inland with most of the population, leaving a defence force composed of roughly 570 militiamen and 200 soldiers stationed at the fort. He also hoped to rely on 400 indigenous archers and 300 galley slaves, but both groups abandoned the defence upon the arrival of Drake's forces. After a brief but intense battle, the defenders retreated, and the English took control of the city. As in Santo Domingo, widespread looting began, followed by negotiations for ransom. Although Drake initially demanded 400,000 pesos, he settled for 107,000. In late February, he convened a council of war with his captains to determine the next objective. With only about 700 soldiers fit for duty due to disease and combat losses, the fleet could no longer hope to assault Nombre de Dios, the expedition's primary target. Instead, they opted to attempt an attack on Havana, where large fleets loaded with treasure and other valuable goods – highly prized in the London markets – regularly gathered. The English fleet reached Cuba in early May, stopping to resupply at Cape San Antonio. A few days later, they arrived at Havana but, upon finding it heavily defended, decided not to engage. Governor Gabriel de Luján dispatched a small fleet in pursuit, and the English lost one ship. The expedition then attacked the small Spanish settlement of San Agustín in Florida. The inhabitants fled, but shortly after, Sergeant Major Powell fell into an ambush and was killed along with much of his detachment. In retaliation, the English razed San Agustín to the ground. Finally, the expedition sailed to Roanoke to collect any colonists who wished to return to England. The survivors reached Portsmouth on 7 September. The mood was likely bittersweet. Over one-third of the expedition's soldiers and sailors had died – most from the malaria contracted in Cape Verde. Moreover, the expedition failed to yield the anticipated profits and did not even recover its costs, rendering it an economic failure. Nevertheless, this outcome did not tarnish Drake's reputation. He successfully convinced Queen Elizabeth and her advisors that the expedition had been a military triumph, despite the setback at Santa Cruz de La Palma. Two major cities – Santo Domingo and Cartagena de Indias – had been severely damaged, and San Agustín had been destroyed. More importantly, the campaign exposed the vulnerability of many Spanish colonial defences throughout the Caribbean. It is also worth emphasising Drake's tactical approach: he consistently targeted undefended cities – preferably by surprise – and unarmed merchant ships. As seen throughout his Caribbean campaign, he avoided engaging heavily fortified cities such as Havana, or treasure ships escorted by war galleons, namely the Spanish treasure fleet. This raises critical questions about the oft-claimed naval superiority of the English at the time.

The Expedition of Thomas Cavendish

On 21 July 1586, Thomas Cavendish set sail from Plymouth at the head of an expedition consisting of three ships, with the aim of completing the second circumnavigation of the globe by English vessels. Seeking to emulate the feat of Francis Drake, he obtained royal authorisation from Queen Elizabeth I to engage in privateering throughout his voyage. Nevertheless, despite his close association with Sir Walter Raleigh and Sir Richard Grenville – whom he had accompanied on the voyage to the Roanoke colony – Cavendish failed to attract the interest of London investors. Consequently, he was compelled to mortgage a significant portion of his estate in order to lease and outfit three ships: his flagship the *Desire* (140 tons), the *Content* (60 tons), and the *Hugh Gallant* (40 tons). His first destination was the coast of Sierra Leone, where he waited in vain for the passage of a Portuguese merchant vessel. As a result, he was only able to replenish food supplies and fresh water. Making use of the Guinea Current, he then crossed to Brazil in mid-November and, by mid-January 1587, began navigating the Strait of Magellan, reaching the vicinity of the settlement of Nombre de Jesús. This town, founded in 1584 by Pedro Sarmiento de Gamboa,[4] was at that time the southernmost inhabited settlement in the world, with a population of 245.

A portion of the settlers relocated in search of more fertile lands and founded the town of Ciudad Rey Felipe. However, extreme weather conditions and the scarcity of food led to the majority of the colonists dying of starvation or disease. By the time Cavendish arrived in the area, only twenty-three survivors remained. Upon the arrival of the three English ships, just three Spaniards came forward to receive them. One of them, named Tomé Hernández, was taken aboard and could be said to have been rescued. The other two Spaniards were left to their fate. Hernández guided Cavendish through the treacherous Magellan Strait, making Cavendish's the third English expedition to successfully navigate it, following those of Drake and Oxenham. Along the Pacific coast, Cavendish devoted himself to raiding and burning small settlements and merchant vessels, acquiring only modest spoils. However, during several of these assaults, he encountered armed resistance from the local population, resulting in significant casualties for the expedition. For instance, in Guayaquil he lost twenty-five men. The losses were such that shortly thereafter, he was forced to abandon the *Hugh Gallant* due to a lack of crew. In July, he captured a small merchant vessel whose pilot, a Frenchman named Michel Sancius, was tortured until he revealed the expected arrival of a galleon from Manila. Cavendish spent the following months patrolling uninhabited coastal regions of present-day Mexico, awaiting the galleon's appearance. At last, fortune favoured him in mid-November 1587, when the *Santa Ana* appeared off Cabo San Lucas, in Baja California. It was an impressive 700-ton vessel, unarmed and heavily laden with goods from China and the Philippines. For four hours, the defenceless galleon was bombarded until, taking on water from multiple breaches, its captain, Tomás de Alzola, surrendered. The spoils were so abundant that it proved impossible to load everything onto the *Content* and the *Desire*; only the most valuable items for the English market were selected – silks, musks, damasks, spices, wines, and, of course, gold. The booty was estimated at approximately 2,000,000 pesos – a truly immense sum that exceeded by more than tenfold the profits obtained by Drake's 1585–6 Caribbean expedition. The passengers and crew of the *Santa Ana* were abandoned on a beach near Cabo San Lucas, and the galleon was set ablaze. Cavendish also seized several Filipino sailors and a Japanese pilot who guided the fleet across the Pacific. They ultimately arrived in Plymouth in September 1588. Cavendish was received by the Queen and subsequently knighted.[5]

The Cadiz Raid of 1587

On 29 April and 10 May 1587,[6] an English fleet under the command of Francis Drake ravaged the port of Cádiz, destroying many ships and supplies destined for the Armada. Subsequently, it inflicted significant damage in southern Portugal, allegedly delaying the departure of the Armada by more than a year. At least, that is the version provided by Drake himself upon his return to London, and which has been uncritically accepted by Anglo-Saxon historiography. On the other hand, Philip II wrote to his ambassador in Paris, Bernardino de Mendoza: 'They [the English] . . . had entered the port of Cadiz as you said it might do, on the same day as, or a little before, your intelligence reached here. The damage it committed there was not great, but the daring of the attempt was so.'[7] So, who was lying, Drake or Philip II? Was Drake's attack truly as successful as he claimed? Did it really delay the assault on the English coasts? One of the key factors in determining whether Drake's expedition was a success, or a failure is to assess whether it fulfilled the objectives outlined by Walsingham in his letter of 21 April to Sir Edward Stafford:

> Drake is gone forth to the seas with four of her Majesty's ships and two pinnaces, and between twenty and thirty merchant ships. His commission is to impeach the joining together of the King of Spain's fleet out of their several ports, to keep victuals from them, to follow them in

case they should be come forward towards England or Ireland, and to cut off as many of them as he could, and impeach their landing, as also to set upon such as should either come out of the West or East Indies into Spain, or go out of Spain thither.

Despite the considerable damage inflicted, none of the expedition's objectives were achieved. As we shall see, most of the losses were sustained by private merchants, who had no involvement whatsoever with the Armada being assembled by the Marquis of Santa Cruz. After an 18-day voyage, a fleet of approximately twenty-three English ships[8] arrived in the Bay of Cádiz at dusk on 29 April. It appears they approached a *bandiere calate*[9] (with lowered flags), prompting Pedro de Acuña[10] to move toward the English formation with a galley and a galliot. As he drew near, the lead ship – *Elizabeth Bonaventure* – opened fire with a broadside that forced the two Spanish vessels to retreat. The cannon fire sparked panic among civilians, triggering a stampede of people seeking shelter within the city walls. Meanwhile, soldiers and militiamen positioned themselves along the bastions, and the English ships anchored opposite the city. Acuña then launched a harassment operation against the English rearguard with his five galleys. He succeeded in boarding a caravel and capturing five English sailors, who confessed that their mission was to take the city of Cádiz. Was that truly their objective? On this point – which is crucial to assessing the success or failure of Drake's raid – there is little clarity. For instance, Corbett,[11] a historian whose writings on Sir Francis Drake verge on hagiography, underscores the lack of planning that characterised the raid. He notes the presence of contradictory orders, such as Lord Howard's unofficial request that Drake not disembark troops (raising the question of whether his official orders were the opposite), even though the expedition included between eight and ten companies under the command of Lieutenant-General Anthony Piatt, amounting to roughly 4,000 men. Upon arriving in Cádiz on 29 April, Drake attempted to capture the city. Securing the Zuazo Bridge – which connected the peninsula to the isthmus of Cádiz – was essential to this objective. By seizing the bridge, the city could be isolated, preventing reinforcements from reaching it from Andalusia. Some of Drake's ships headed in that direction, only to find two of Acuña's galleys anchored there. They were careening, so it means that they were without crew or artillery. Nevertheless, the English vessels turned back without firing a single shot, raising doubts about the combat-readiness of certain ships in the expedition. This episode is typically overlooked by Anglo-Saxon historiography.[12] It is key to understanding that Drake's main plan had just failed and how, in a fit of anger, he then decided to inflict as much damage as possible on the ships anchored in the harbour. On this point, the authors Tanturri (2012), Ridella et al. (2016), and Gómez Beltrán (2022) agree. Beyond attempting a surprise seizure of the city, the renowned English corsair had no concrete plan for its conquest. As was the case in his raids in the Caribbean, he only targeted cities that could be taken by surprise or that had minimal or non-existent defences. We might recall his previous raid and what occurred in places such as Cartagena de Indias or Havana, for example. Does this mean that Drake was a coward? Certainly not. As previously discussed, – and this episode serves as a clear example – Drake was an extraordinary privateer, arguably the greatest of them all, but he did not possess a military strategist's mindset. Following this failure, the English ships began to massacre the vessels anchored in the port. The first to be sunk was a Genoese ship of 600 tons, the *San Giorgio e Sant'Elmo*, which was laden with jars of olives, barrels of cochineal, boxes of agricultural produce, a variety of ceramic vessels, tropical woods (lignum vitae), dividers for navigation, leather shoe soles[13] . . . worth 40,000 ducats, which had attempted to flee in haste.[14] Meanwhile, in his palace at Sanlúcar de Barrameda, the Duke of Medina Sidonia, Captain General of Andalusia, was celebrating the birth of his daughter. As soon as news of the English attack reached him, he organised reinforcements to secure the Zuazo Bridge and the defence of Cádiz, dispatching the first units of cavalry and infantry. Although their mission was to defend the city, the anchored ships were left at the mercy of the English, despite the efforts of Pedro de Acuña's seven galleys, which were already fully engaged in repelling repeated assaults. At the same time, messengers departed from the city to inform Philip II, who was in Aranjuez, and to Lisbon to alert the Viceroy of Portugal, Archduke Albert VII of Austria, the King's nephew, as well as the Marquis of Santa Cruz. Over the following days, according to Spanish reports, twenty-four vessels were burned and sunk, while another four were captured. Among those destroyed was the galleon *Nuestra Señora de la Concepción de Guadalupe*, an 800-ton ship owned by the Marquis of Santa Cruz, and a nao owned by Esteban de Murrieta loaded with iron. From the English perspective, the raid had been a resounding success. Drake submitted to Walsingham[15] a report – somewhat exaggerated – listing the ships he had sunk and the wide array of provisions he had captured or destroyed, which supposedly dismantled Philip II's logistical preparations for the invasion of England. This report became the foundation of British historiography in evaluating Drake's raid. However, the inventory of damage received by Philip II, though significant, includes details that must be considered in assessing the real impact of Drake's attack on the Armada's logistics: 3,443 quintals of hardtack, 3,288 fanegas of wheat, 392 barrels and casks, 200 quintals of iron, and 200 muskets were either destroyed or captured, in addition to the merchandise aboard the merchant ships anchored in Cádiz. In total, the damages amounted to 172,000 ducats – approximately £750,000 (equivalent to £128,735,475 at current prices[16]) – but English authors, such as Hutchinson,[17] fail to mention that only 17,426 ducats (£13,038,209 at current prices[18]) of this sum belonged to Philip II's Treasury, while the rest were owned by private merchants.[19] Did the Armada's logistics depend on these supplies? No. To provide a reference for the damage inflicted on the Armada's total provisions, Drake destroyed 3,443 quintals of biscuit, whereas the Armada carried 96,155. Regarding the 200 muskets captured, Medina Sidonia had already sent about 1,000 with the first reinforcements that arrived in Cádiz. That covers supplies, but what about the ships? Were the twenty-four ships sunk and the four captured intended for the Armada? Once again, no. Among those sunk and destroyed, most (sixteen vessels) were hulks or carracks with no combat utility or were French. The remainder included four naos from the Indies Fleet, three of which were small and unsuitable for inclusion in the Armada. Only Diego Lorenzo's nao, 500 tons, could potentially have been part of the Enterprise of England. The other two larger vessels were previously mentioned: the galleon of the Marquis of Santa Cruz and Mugarrieta's nao.[20] Thus, out of the total twenty-eight ships sunk or captured, only three could have sailed to England. The Armada was scheduled to depart in September, and Drake's raid did not alter any of its plans. Gómez Beltrán concludes[21] that not only did Drake fail to disrupt the Armada's logistics, but he also failed in his primary objective: capturing Cádiz. As a result, he had constant disputes with his vice-admiral, William Borough – a seasoned sailor with an outstanding record in the Royal Navy – who consistently criticised Drake's lack of a pre-established plan and his authoritarian command style during the operation. Borough later expressed these concerns before the Admiralty:

> but bare away for the place presently without consultation or order given to the fleet, in such confused order as was never heard of in such an action . . . but if a man of government and judgment such as had been fit for the action, that would have heard men's opinions and dealt advisedly and with order, had had the direction and handling of it, we might as well have had the spoil of the galleys and town of Cadiz with

the rest of the shipping we left behind us in that Bay, as those shipping that we had thence and spoiled in that place.[22]

That is to say, beyond the destruction of a significant portion of the ships anchored in Cádiz, the lack of a clear objective and poor execution prevented the damage inflicted on Philip II from being truly significant. On 2 May, the Bay of Cádiz awoke without the threatening silhouettes of the English ships. Unaware of Drake's plans, warnings were sent to the Canary Islands and major Caribbean cities in case the English corsair intended to continue his raid in those territories. However, his goal was to remain at Cape St. Vincent, on Portugal's southern coast, to prevent new Spanish ships from joining the fleet assembling in Lisbon. There, Drake decided to disembark and seize the city of Lagos, which once again brought him into conflict with Borough. The operation took place on 4 May, with around 1,000 men going ashore, but they did nothing more than parade in front of the city walls. These walls, defended by a minimal garrison and a few militiamen, were enough to deter the English troops from attacking, forcing them to re-embark and casting further doubt on their combat capabilities and lack of strategic planning. Days later, Drake turned his attention to Sagres Castle, which he captured with virtually no resistance after its small garrison abandoned it upon witnessing the overwhelming superiority of the English infantry. Drake's fleet spent the following weeks along Portugal's southern coast, burning some towns and attacking small fishing boats – all actions with no strategic value. By this point, a desperate Borough – highly critical of the expedition leader – had been dismissed by Drake and confined aboard the *Golden Lion*. And what did the Spaniards do during this time? Philip II's admirals were reassured upon realising that Drake was not continuing his expedition to the Caribbean and, to some extent, allowed the English corsair to carry on with his futile raids along Portugal's coast, where the damage was minimal. Meanwhile, he was monitored by two squadrons: one from Lisbon under Alonso de Bazán's command, with seven galleys; and another from Cádiz under Martín de Padilla's command, with ten galleys – both of which Drake avoided engaging at all costs. At no point did Drake dare approach Portugal's capital. By early June, English sailors began showing signs of exhaustion and frustration, prompting Drake to send part of his fleet back to England. In his reports to England (which many Anglo-Saxon historians judge uncritically), Drake claimed to have burned barrels with a storage capacity of 25,000 tons – equivalent to about 22,000 English barrels or roughly 40,000 pipes destroyed. This would have required no fewer than 400 vessels of 60 English tons each – an exaggeration that William Borough himself contradicted in a letter to the Lord High Admiral:[23]

> Since which time we have taken about 28 or 30 ships and barks, whereof one was a flyboat of Dunkirk with merchandise of great value in her, bound for St. Lucar, another a small hulk of Holland, laden with timber, Spaniard's goods, which came from Galicia, bound also for St. Lucar, both which we possess and keep by us. The rest were small carvels and barks of burden between 16 and 40 tons, most of them laden with pipe-boards, hoops, timber, rafters for oars and such like lading of small value, and some had nothing but ballast.

Drake's next step was to head to the Azores, where he hoped to encounter a ship arriving from America. A storm dispersed his fleet, which allowed the crew of the *Golden Lion* to desert and return to England. Drake achieved his prize on 18 June, near the island of São Miguel, where he intercepted one of those Portuguese ships loaded with treasures that typically sailed alone, without the protection of the Indies Fleets. The vessel was the carrack *São Felipe*, whose cargo was valued at £108,049 (equivalent to £19,567,792[24]). With his loot secured, Drake decided to return to England, docking at Plymouth on 28 June. The reception must have been euphoric. Active English propaganda boldly claimed that Drake had 'singed the King of Spain's beard', a mantra repeated uncritically ever since. However, once again, the losses are overlooked. It has been documented that sixteen of the twenty-three ships returned, many with their crews decimated by an epidemic.[25]

To determine whether the raid was as successful as Drake boasted, one must answer a key question: Did it delay Philip II's plans? The answer is no. Undoubtedly, capturing the *São Felipe* was an economic success but a Pyrrhic victory considering that Drake failed to attack the Indies Fleet – a mission explicitly outlined in Walsingham's letter dated 21 April. Furthermore, Drake's reports exaggerated figures (as even his lieutenant refuted), and later authors such as Hutchinson[26] or Parker & Martin[27] failed to distinguish between direct damage inflicted on Philip II's Armada and that affecting private merchants – information verifiable in reports received by the Spanish king,[28] intended for private use rather than propaganda purposes. Nonetheless, a second question arises: Was the Armada's departure delayed? Yes, it was indeed delayed – its departure had been scheduled for mid-September – but not due to Drake's raid; rather, it was caused by a combination of factors that will be explored later.

Leicester in the Low Countries

Robert Dudley, Earl of Leicester, disembarked at the port of Flushing on 20 December 1585 (10 December in the English calendar of the time), in response to the Treaty of Nonsuch. The Queen, who had rejected sovereignty over the Netherlands in May, nevertheless committed to increasing her support for the Dutch rebels' cause: England would provide military aid to the United Provinces.[29] Various logistical issues delayed the arrival of the English expeditionary army, composed of approximately 7,000 men, until December. Among them were several nobles and relatives of Leicester, such as his nephew the renowned poet Philip Sidney, and his brother-in-law, William Knollys. Dudley was received as a saviour of the nation, celebrated in cities like Amsterdam and The Hague with festivities, banquets, and gifts. In early January, he was offered the position of Governor-General of the Netherlands. Historians often claim that communications with England were cut off for two months due to bad weather; however, by mid-January Dudley had already accepted the position, with his official proclamation occurring at the end of the month.[30] It is peculiar how Leicester allowed himself to be pressured or convinced into accepting this role against his Queen's orders, which had only granted him command over her army and instructed him to advise the Dutch on political and military matters. Dudley's decision enraged Elizabeth[31] – though perhaps not as much as Philip II. On 5 December, Philip was in Binéfar (Valencia) when he received a letter from Parma detailing preparations in Flushing to welcome the English expeditionary army. The Duke expressed his frustration, as he had hoped that the fall of Antwerp would open an opportunity for peace negotiations with the Union of Utrecht, which some of its leaders were beginning to demand. Nevertheless, he continued his campaign to seize Friesland and northern Brabant. Meanwhile, Dudley wasted no time organising his army. He appointed trusted individuals as governors of strategic cities: Peregrine Bertie, Lord Willoughby de Eresby in Bergen-op-Zoom; Sir John Conway in Ostend; and Sir Philip Sidney in the strategic port of Flushing. In January 1586, Colonel Francisco Verdugo triumphed at the Battle of Boksum in Friesland, crushing an army composed of mercenaries and Frisian volunteers. Parma had to wait until April to begin his campaign in northern Brabant, starting with the siege of Grave under Pedro Ernesto de Mansfeld's command. Given Grave's importance, Dudley sent part of his army to relieve the city. The first skirmish between English and Spanish troops occurred here; Dudley's forces won by capturing Batenburg's small fortification.[32] Mansfeld requested reinforcements from his son-in-law Colonel Verdugo but had to leave for Groningen due to his wife Dorotea Mansfeld's unexpected death. Despite this interruption, Spanish artillery began battering Grave's walls with twenty-three cannons. In early June, Farnese personally organised the assault. Grave's governor, Baron Peter van Hemart – seeing no external support forthcoming – negotiated surrender on 6 June. Shortly thereafter, rebel-controlled fortifications at Megen and Batenburg also fell. Furious at Hemart's actions, Dudley ordered his execution – a decision that shocked the Dutch and led them to lose faith in him as their leader. Farnese's next target was Venlo, which he quickly besieged. Relief forces led by Maarten Schenck and Roger Williams launched a surprise attack on the Spanish camp at dawn but were repelled with heavy casualties. Venlo surrendered on 28 June. Shortly after this victory, Philip II ordered Parma to intervene in Cologne's war; Farnese moved much of his army to besiege Neuss, which fell within days. In response, Leicester ordered a siege on Zutphen. Colonel Francisco Verdugo sent a convoy carrying weapons, food supplies and soldiers but was intercepted by Leicester's army during a skirmish on 22 September 1586. The convoy was led by Italian-Albanian cavalry under Captain George Cresiac. Initially, English cavalry – including prominent nobles like William Russell (Baron Russell of Thornhaugh), Peregrine Bertie (Baron Willoughby de Eresby), Sir Francis Vere, and Sir Philip Sidney – overpowered Cresiac's forces and captured him. However, Spanish tercios followed closely behind; their deadly musket fire routed Leicester's troops and forced them to lift Zutphen's siege. This skirmish might have gone unnoticed in history books if not for one significant event: Sir Philip Sidney – Leicester's nephew and a distinguished young writer – was wounded by musket fire while leading a final desperate cavalry charge. He succumbed to gangrene some days later.

After failures on the battlefield and a highly questionable political and economic administration, Leicester returned to England in December 1586. Shortly thereafter, two senior commanders of his army, Rowland York and William Stanley, defected to the Spanish side along with a significant portion of their troops; moreover, York handed over the town of Deventer. The leaders of the United Provinces decided to dismiss Leicester as Governor-General, appointing Maurice of Nassau in his place, while Lord Willoughby de Eresby assumed command of the English expeditionary army. However, Leicester returned in June. Once again leading his army, he directed his troops to the strategic port of Sluis, which was under siege by the Duke of Parma. On 19 July, Leicester arrived near the city with 4,000 soldiers and approximately 400 cavalrymen but did not dare attack the Spanish forces, whose numbers were half his own. Leicester had been expecting reinforcements from the United Provinces, which never arrived. Stripped of his title as Governor-General, he no longer held any authority over Dutch soldiers. Consequently, he withdrew from Sluis, much to the despair of its governor Arnold de Groenevelt and the English garrison led by the tireless Roger Williams. With no hope of external assistance and breaches in its walls, Sluis's governor negotiated an honourable surrender – a move Leicester denounced as treason.[33] The situation for the English did not improve in subsequent months as they failed in their attempts to capture Leiden, Enkhuizen, and Hoorn. To further complicate matters, auditors were sent from England to review Leicester's questionable financial accounts. One of these auditors, writer George Whetstone, accused Sir Edmund Uvedale (or Udall, depending on the source) of embezzlement. This led to a duel in Bergen-op-Zoom that resulted in Whetstone's death. In October 1587, Leicester returned permanently to England with his reputation in ruins and financially bankrupt. That same year, Elizabeth I began discreet peace negotiations with the Duke of Parma through André de Loo. As Van Der Essen sharply observes,[34] England's economic situation was dire following the loss of trade with Spain and Portugal but especially due to Parma's troops controlling the mouth of the Rhine – cutting off English merchants' access to German territories and parts of Flanders. Nevertheless, Philip II never trusted Elizabeth's true intentions and was unwilling to sign a peace agreement after Drake's raid on Cádiz in spring 1587 – not because of the damage caused (as discussed earlier) but due to the audacity of directly attacking peninsular territory. Despite this, peace talks continued; for instance, in April 1588 Philip II sent Parma his conditions for reaching an agreement:

> The first demand was that free use and practice of our holy Catholic faith be allowed in England for all Catholics, both native and foreign, lifting the exiles of those who had been banished from the kingdom. The second was the restitution of the territories in the Netherlands currently occupied by my forces. The third was compensation for the damages inflicted upon me, my states, and my subjects, which amounted to an excessive value.[35]

These demands could be summarised as a restoration of the status quo that existed prior to England's provocations against the Spanish Crown. The negotiations were followed by gestures of détente. For instance, that same month, the Queen allowed forty German ships detained in English ports, loaded with goods destined for Spain,[36] to depart. However, while negotiations continued, both countries also prepared for war. In fact, letters from the Duke of Parma to the King of Spain often began by detailing the progress of peace talks but then proceeded to describe preparations for the invasion of England.[37] Similarly, Spanish intelligence received regular updates on England's war preparations.[38] It is true that, based on contemporary documentation, both sides sought an understanding. However, it is equally true that they approached negotiations from maximalist positions and seemed united only by an unshakable mutual distrust.[39] As Van Der Essen[40] notes, by mid-June, following the English envoys' final refusal to accept Spanish conditions and – importantly for Elizabeth – the refusal of the United Provinces to participate in (and ratify) any agreements, negotiations collapsed.[41] At that point, nothing could stop the war. Nothing could stop the Enterprise of England.

Chapter 4

THE ENTERPRISE OF ENGLAND

It looks like we have a plan

By the late 1560s, Pope Pius V was fervently determined to restore Catholic supremacy in England, and thus Elizabeth I, as Supreme Governor of the Church of England, became one of his primary adversaries. Intent on deposing her and replacing her with Mary Stuart, he sought assistance from Philip II through his ambassador in Rome, Juan de Zúñiga y Requesens.[1] However, the King of Spain preferred to see the English question resolved through diplomatic means, as his principal concern at the time was the Ottoman Empire. Pius, who was concurrently orchestrating the alliance among the Mediterranean Catholic powers that would culminate in the Holy League's victory at the Battle of Lepanto (1571), had to content himself with excommunicating Elizabeth through the papal bull *Regnans in Excelsis* ('Reigning on High'), dated 27 April 1570. His successor, Gregory XIII, was no less uncompromising toward the Queen. He even went so far as to organise two expeditions to Ireland (1578 and 1579), both of which failed. At the request of the English Jesuits, however, he later moderated his predecessor's bull by advising English Catholics to outwardly obey the Queen in civil matters. The next pope, Sixtus V, was equally persistent in his opposition to Elizabeth. He attempted to form an alliance with the Duke of Guise and Philip II to invade England – an initiative that the King of Spain opposed.

In Spain, the first to propose the invasion of England was the Marquis of Bazán, following the Battle of Terceira Island in August 1583. However, there is some ambiguity surrounding his report, which was not a concrete invasion plan but rather a list of the resources that would be required to carry it out. In any case, Philip II did not seriously consider the invasion of England until late 1585, specifically on 11 November, when he received a letter from the Duke of Parma informing him of the arrival of the first contingent of English troops in Flanders. In his response to his nephew, the King requested the drafting of a plan to land Spanish forces in the realm of Elizabeth I. A few days later, in Lisbon, the Marquis of Santa Cruz learned of the King's intentions and offered to prepare a separate plan – an offer that was accepted. By late March, Philip II received Bazán's proposal, which, drawing on the successful experience at Terceira, called for an amphibious landing. However, the plan was Homeric in scale, as it required 'no less than 510 ships, including 150 galleons and other warships, 40 heavy transports, and 320 auxiliary craft, 30,332 seamen to man them, and 63,890 soldiers, including 1,200 cavalry with mounts, 4,290 artillery men, and 55,000 Spanish, Italian, and German infantry. The total cost he calculated at nearly four million ducats.'[2] The estimated cost is 3,800,000 ducats.[3] Shortly thereafter, the King received Parma's plan. It involved the transport of 30,500 soldiers from Flanders to England using approximately 200 landing barges and 20 escort ships, carried out in less than 12 hours in a surprise attack. For this, three indispensable conditions were required: maintaining the plan's secrecy, ensuring France's non-involvement (given the interest shown by the Duke of Guise), and securing the military situation in the Low Countries. It is often believed that the final outcome resulted from Philip II merging these two plans, but the reality is very different and far more complex. It is true that both plans were studied at court during a War Council attended by the King, Juan de Zúñiga Juan de Idiáquez, secretary to Philip II;, Cristóbal de Moura, advisor on Portuguese affairs, and the Count of Chinchón. Ultimately, Bernardino de Escalante[4] was tasked with presenting the operational plan, which he submitted on 26 July 1586. His plan could be divided into three phases: first, an Armada would land in Ireland as a diversionary manoeuvre to draw off part of the English fleet and army; second, a Catholic uprising would be promoted in England; and finally, a direct attack from Flanders would be launched. Following his recommendations, the Marquis of Santa Cruz was appointed Captain General for the troops destined for Ireland, with Lisbon designated as the base from which the Armada would depart and begin gathering ships, men, weapons, and provisions for the Enterprise. The movement of ships soon attracted the attention of diplomats and spies who, although unclear about the ultimate objective of this formidable Armada being organised, nonetheless alerted the English court. In response, they decided to send Drake to Cádiz to disrupt Spanish plans. By April 1587, the effective force assembled consisted of Portuguese galleons, Recalde's squadron, and Antonio Hurtado's ships. However, Oquendo's squadron (being organised in Cantabria) and those from Naples and Sicily had yet to arrive. Paradoxically, Drake's actions did not so much delay the Armada as activate it. The fear that he might attack the Indies Fleet prompted urgent mobilisation of many ships stationed at the mouth of the Tagus River.[5] Throughout that summer, sightings of groups of English ships continued along Portugal's coast – for instance, some ships from Drake's expedition that had been scattered after early June storms were likely those spotted near Madeira weeks later. Spanish authorities feared they were preparing an attack on the Indies Fleet and quickly began organising an escort fleet. Matters escalated when news arrived that Corvo Island (the smallest in the Azores archipelago) and Vila das Lajes on Flores Island had been devastated by English corsairs.[6]

Additionally, spies in England reported the concentration of a fleet at the mouth of the Thames,[7] as appears to be confirmed by a letter from Walsingham to Burghley dated 16 July, in which he states: 'urged that Drake should return to the Azores to attack the lumbering wide-bellied treasure ships bringing back bullion from the Spanish empire in the Americas. The best way to bridle their malice is the interruption of the Indian fleets'.[8] Around this time, the Ocean Sea Armada under the command of the Marquis of Santa Cruz departed Lisbon with 36 ships to escort the 107 vessels of the Indies Fleet, while another fleet under Martín de Padilla patrolled the Portuguese coast. However, Drake did not set sail. Here, sources contradict each other. Hutchison claims that the operation was cancelled by the Queen, who still hoped for fruitful peace negotiations with the Duke of Parma. In contrast, Gómez Beltrán argues[9] that it was due to knowledge of Spain's massive escort force. The same author reports that during these weeks, four pirate ships – three French and one English – were captured. Additionally, it is reported that the Portuguese *Santa María*[10] was rescued while being attacked by two English ships and two vessels. Meanwhile, logistical challenges continued to plague the organisation of the Armada, as cannon and artillery crews were not yet fully supplied. Furthermore, no port in Flanders had been captured that was large enough to allow Parma to transfer his troops to England. By late August, the Indies Fleet contacted Bazán's escort fleet, which arrived at the mouth of the Guadalquivir on 25 September and carried out the unloading of American treasure amidst a terrible storm. The good news for Philip II was that all the cargo – valued at 16,000,000 ducats (£449,077,238 in current prices) – was successfully secured.[11] A few days earlier, on 14 September, the final plan for the Armada's landing in England was established. It ruled out an initial attack on Ireland and decided instead that, rather than a war fleet as proposed by Santa Cruz, the Armada would be a troop transport convoy escorted by galleons tasked with joining Parma's prepared fleet. At this point, everything seemed ready for departure;[12] only the Ocean Sea Armada and Oquendo's squadron were yet to arrive. Moreover, news of Sluis's capture provided Parma with a new strategic advantage. Philip II was eager for the invasion to proceed but an unforeseen issue disrupted plans: a severe storm struck Santa Cruz's Ocean Sea Armada, delaying its arrival in Lisbon until early October. Although all the ships managed to reach Lisbon, many suffered severe damage requiring weeks of repairs. Meanwhile, Oquendo's squadron had encountered another storm *en route* to Lisbon, breaking the bowsprit and foremast of its flagship and forcing repairs at Laredo (Asturias). His squadron did not reach Lisbon until 28 October. On 16 November, yet another storm in Lisbon caused further damage to numerous ships. Thus, it was Aeolus – the god of winds – and not Drake who prevented the Armada's departure on its scheduled date in October 1587.[13] The King set a new departure date for late November, but Santa Cruz continued to face logistical problems and deemed winter an entirely unsuitable time for the expedition. In addition to the adverse weather conditions – storms and dense fog – he believed that the limited daylight hours were insufficient for carrying out a landing operation in hostile and unfamiliar territory with such a large force and so few guarantees of success. By that time, the Marquis had grown increasingly sceptical of the mission's chances, a pessimism that became evident in his correspondence. The King, in turn, reproached him for his attitude, a rebuke that deeply wounded the veteran admiral. Meanwhile, the situation in Flanders was deteriorating. The Duke of Parma grew desperate over the delays of the fleet. An epidemic was ravaging his army, particularly the German contingents, many of whom began deserting en masse. Parma also demanded more money and supplies from his uncle to sustain an army of such magnitude. Shortly thereafter, he wrote to inform Philip II of the landing of English troops in Flanders. In December, King Philip – by then 60 years old and suffering from various chronic ailments – fell gravely ill. This temporarily halted preparations for several weeks, as all matters relating to the English campaign required his direct oversight. Upon his recovery, he ordered the division of the Armada into two contingents: the first was to set sail immediately, carrying the reinforcements Parma needed to complete his army; the second, scheduled to depart later, was to ensure the success of the eventual landing in England. Once again, however, adverse weather conditions made departure impossible. It was decided that the new date for the operation would be 1 February 1588. Yet by that time, it was Parma who admitted that he was not ready and that he would need to conquer the city of Flushing first – its port being crucial for launching an invasion of England. This revelation irritated Philip II, who nonetheless, faced with the unavoidable delay, instructed his nephew to resume peace negotiations with Elizabeth. From Lisbon came further disheartening news: a typhus epidemic had begun to decimate the soldiers and sailors, many of whom had been aboard ships for months. In mid-January 1588, the court dispatched the Count of Fuentes[14] to resolve the persistent logistical problems that the Marquis of Santa Cruz seemed unable to overcome. But was the blame solely Santa Cruz's? According to Gómez Beltrán, it was not.[15]

The admiral had not been granted full authority over the organisation of the Armada; many decisions were imposed from Madrid, where a tangle of bureaucracy and indecision delayed the fleet's departure. To make matters worse, the typhus epidemic reached Bazán, whose health rapidly deteriorated. On 9 February 1588, the Marquis of Santa Cruz passed away. With his death, Spain not only lost arguably the greatest admiral in its history – having served the Crown for nearly 50 years, fought across various seas against multiple enemies, and never suffered defeat, an unmatched record – but the Armada also lost perhaps the only commander capable of leading it to success. His position was temporarily assumed by his brother, Alonso de Bazán. At El Escorial, names were quickly put forward to take permanent command of the Armada. Several candidates were considered – Martínez de Leyva or Recalde, for instance, both of whom would later participate in the campaign, as well as the Count of Fuentes, who reportedly felt relieved not to be chosen.[16] However, the King and his advisors agreed that what was required was not merely a military leader, but an effective administrator – someone capable of 'launching the ships to sea' – and who also possessed a noble title of sufficient prestige. The most suitable candidate, without question, was Alonso Pérez de Guzmán, the Duke of Medina Sidonia. His principal drawback – his lack of naval warfare experience – was to be compensated by a secondary command structure consisting of seasoned admirals such as Recalde, Oquendo, Bertendona and Flores de Valdés, along with Martínez de Leyva. For navigation, he would rely on the advice of Captain Marolín de Juan, pilot of the *San Martín* and Pilot-General of the Armada. Initially, however, the duke refused the appointment. In a now-famous letter addressed to Juan de Idiáquez on 14 February,[17] Medina Sidonia attempted to decline the command, citing health concerns[18] and the dire financial condition of his household – he claimed to be in debt to the sum of 900,000 ducats. Yet the King's persistence eventually overcame the duke's reluctance, and in a letter dated 26 February, Medina Sidonia formally accepted the position.[19] His appointment was well received by foreign ambassadors and even by the Duke of Parma.[20] On 14 March, Medina Sidonia arrived in Lisbon. Once in the Portuguese capital, he immediately set to work, displaying his most distinguished virtue: exceptional organisational capacity. He was briefed on the status of the ships, the sailors and soldiers, as well as the artillery, ammunition, and provisions required by the Armada. The preparations were already significantly advanced,[21] and his directives primarily consisted of reducing the weight of certain galleons to improve manoeuvrability and speed and redistributing artillery among the vessels to balance the firepower between the various squadrons[22] – particularly those under Oquendo and Diego Flores. He also had to confront the shortage of sailors, especially skilled pilots familiar with navigating the English Channel and the Flemish coast. This issue was partially mitigated by the arrival of pilots from Flanders, who distributed navigational charts to the principal ships of the Armada. Lastly, Medina Sidonia addressed the shortfall of certain essential foodstuffs, such as cheese and fish, which were partially compensated by a greater quantity of hardtack than originally anticipated. As for ammunition, while the Marquis of Santa Cruz had stipulated that each cannon should be equipped with a minimum of thirty cannonballs, Medina Sidonia ordered this number raised to fifty – an ambitious target few cannons would reach, though all departed with at least the minimum set by the Marquis. Perhaps Medina Sidonia's greatest achievement, however, was changing the morale of the Armada. The intense pressure that Philip II had placed on Santa Cruz – and, by extension, on the rest of those responsible in Lisbon for assembling the fleet – without offering timely solutions, had plunged the Marquis into a state of gloom that affected his collaborators as well. Medina Sidonia's arrival brought renewed energy to the preparations, which, although already advanced, benefited greatly from the duke's efficient and resolute character. Within a matter of weeks, the fleets were ready to depart. By early May, all that remained was for Aeolus to grant permission to set sail.

As for the Duke of Parma, preparations were also proceeding according to plan, though not without significant difficulties. Farnese had to contend with two fundamental challenges: first, identifying a suitable port from which to embark his troops, and second – and no less critical – recruiting the necessary forces for the invasion. At this point, it is worth pausing to examine the troops that had been assembled to invade England. For over 150 years, the Spanish Tercios were the most formidable infantry units in Europe – and arguably, in the world. Their origins can be traced to the early sixteenth-century military reforms initiated by Gonzalo Fernández de Córdoba (1453–1515), known as 'El Gran Capitán' ('The Great Captain'),[23] who introduced the use of firearms within infantry formations. He professionalised military service through the establishment of permanent volunteer units, replacing the ineffective medieval levies and the costly mercenary forces, which had previously represented the closest approximation to a professional army. In addition, he revolutionised infantry tactics. The term Tercios came into formal use following the military ordinances of 1534 and 1536. The first such units were Spanish infantry regiments stationed in Italy: the Tercio of Naples, the Tercio of Sicily, and the Tercio of Lombardy. These were later joined by the Tercios of Sardinia and the Galley Tercios – the latter being the first marine infantry unit in history. Each Tercio comprised approximately 3,000 men under the command of a field master (*maestre de campo*), whose second-in-command was a sergeant major (*sargento mayor*). The Tercios were subdivided into companies of around 300 soldiers, each led by a captain, who was assisted by lieutenants (*alféreces*). Their success lay primarily in their innovative combination of weaponry – pikes, arquebuses, and muskets – along with their tactical flexibility, strict discipline, exceptional morale, and a powerful sense of honour. Moreover, the officer corps was composed of men of outstanding capability. The Battle of Cerignola (1503), fought against the French army led by Louis d'Armagnac, Duke of Nemours (who was killed during the engagement), marked the beginning of the Tercios' undisputed military dominance – a supremacy they would maintain across continental Europe until the Battle of the Dunes in 1658. For generations, serving in the Tercios was a path to glory and prestige for members of the nobility. For the lower social strata of Spanish society – especially Castilians – it offered an opportunity for upward social mobility and an escape from poverty. Interestingly, many of the most important poets, playwrights, and authors of Spain's Golden Age served in the Tercios. Among them were Garcilaso de la Vega (1503–36), regarded as the foremost Spanish Renaissance poet, who died from wounds sustained during the siege of Le Muy[24] while serving as a *maestre de campo*; Lope de Vega (1562–1635);[25] Pedro Calderón de la Barca (1600–81);[26] and, of course, Spain's greatest literary figure, Miguel de Cervantes (1547–1616), who was wounded at the Battle of Lepanto in 1571. Thus, it was logical that the army tasked with deposing Elizabeth I would consist of the Tercios. These units were to be divided into two main corps: those embarked in Lisbon and those prepared by the Duke of Parma in Flanders. In his initial plan, Farnese estimated that he would need around 30,000 men to defeat the English army. At that time, the nephew of the King of Spain had only about 4,300 soldiers, divided among three Tercios: Cristóbal de Mondragón's (twenty-seven companies), Juan de Águila's (twenty-four companies), and Francisco de Bobadilla's (twenty-one companies).[27] However, these numbers were unattainable and were subsequently reduced. Later, under Escalante's plan, when the combined action of an Armada with Parma's army was agreed upon, the required number of troops for Parma was reduced by one-third. Nevertheless, more soldiers were still needed – not only to attack England but also to maintain positions in the Netherlands. Thus, in early February, a levy was conducted to recruit volunteers from the Kingdom of Castile. However, the number of recruits fell short of expectations, leading to the exceptional decision to accept Catalan volunteers. The responsibility for this recruitment fell to Lluís de Queralt, an experienced veteran of Mediterranean wars who volunteered to raise a Tercio composed of soldiers from southern Catalonia.[28] He successfully recruited ten companies, numbering fewer than 1,000 men, most of whom were bandits[29] offered royal pardons in exchange for military service. These soldiers had the advantage of being battle-hardened and militarily experienced – unlike many Castilian peasants who had never handled firearms – but their discipline was highly questionable. By spring 1588, Parma had slightly over 10,000 Spanish soldiers in Flanders. The landing troops also included contingents from nations allied with Spain and even a Tercio composed of English, Scottish, and Irish soldiers. Flanders was a common destination for English Catholics exiled after failed uprisings against Elizabeth I's Protestant regime – for example, Charles Neville, Earl of Westmoreland, who fled England (and later Scotland) after his unsuccessful Catholic rebellion in 1569. In Flanders, they enjoyed Spanish protection while remaining close enough to England to return quickly if political circumstances changed. Among the first British soldiers to enter Spanish service was the Scotsman William Semple,[30] who handed over Lier to Parma in 1582 along with his battalion of Anglo-Scottish soldiers. The following year, Captain Moore and his troops surrendered Aalst to Parma. In January 1587, Colonel Roland York defected at Zutphen Fort with his cavalry and joined the Spanish side. The most significant defection occurred days later when Colonel William Stanley[31] surrendered Deventer – a strategic city – to Parma along with over 1,000 Anglo-Irish soldiers. As one of Leicester's most trusted men, Stanley's defection dealt a severe blow to the morale of the English Expeditionary Corps. The final contingent from the British Isles to join Spain's side came from Geldern under Colonel Archibald Patton, who defected along with his Scottish troops.[32] Parma sought to leverage this substantial number of English, Scottish, and Irish soldiers for the Enterprise of England and offered Stanley command over a Tercio composed of these nationalities – ultimately assembling slightly over 1,000 men ready for embarkation at Dunkirk by summer 1588. Ninety per cent of this Tercio consisted of deserters from the English army in the Low Countries.[33] Over subsequent months, Parma's Spanish Expeditionary Corps for landing operations was completed by approximately 4,000 Neapolitan soldiers under Carlo Spinelli de Castrovillari;[34] 4,250 troops from other Italian regions under Camillo Capizucchi;[35] nearly 1,000 Burgundians led by Marc de Rye de la Pallud[36] (Marquis of Varenbon); and 3,500 Germans under Karl von Habsburg.[37] These were the forces Parma would command for the Enterprise of England.

The Spanish infantry contingent was composed of the following:[38]

- Tercio of Cristóbal de Mondragón: 30 companies with 2,968 men.
- Tercio of Juan Manrique de Lara: 30 companies with 2,663 men.
- Tercio of Francisco de Bobadilla: 23 companies with 2,226 men.
- Tercio of Lluís de Queralt: 9 companies with 861 men.

The contingent from allied nations consisted of:

- Tercio of Stanley, composed of English, Scottish, and Irish soldiers, approximately 1,000 men.
- Regiment of Varenbon, composed of around 1,000 Burgundian soldiers.
- Tercio Spinola, with approximately 1,500 Neapolitan soldiers.
- Tercio Capizucchi, with another 1,500 soldiers from other regions of Italy.
- Walloon regiments, approximately 7,000 soldiers under the command of Philibert. Emmanuel de Lalaing, Marquis of Renty; Pierre de Henin, Count of Boussu; Octave. de Mansfeld; Valentin de Pardieu, Baron de la Motte aux Bois; Philibert de Rye, Count of Varax and Baron of Balançon; and Antoine de Grenet, Seigneur de Werpen.
- German regiments, around 8,000 men under the command of Juan Manrique de Lara and Ferrante Gonzaga.

Regarding the embarkation port, the best option was undoubtedly the port of Flushing. However, it was in Anglo-Dutch hands. Farnese repeatedly requested time (and funds) from Philip II to capture it, but the King – eager to begin the invasion as soon as possible – denied his request. With the mouth of the Scheldt River blocked by an Anglo-Dutch fleet under Captain Wynter's command,[39] Parma had no choice but to transport his men and their flat-bottomed boats through inland canals to the vicinity of the shallow ports of Nieuwpoort and Dunkirk. He hoped that Medina Sidonia's Armada could approach closely enough to escort these flatboats – some so poorly equipped that they lacked sails and would require towing.

Meanwhile, in Lisbon, the wind remained still. With all the men embarked but the ships stalled at the mouth of the Tagus, Medina Sidonia feared a potential shortage of provisions. He sent messengers to the coast to prepare a shipment of food and, anticipating further complications, organised a healthcare system. Physicians were dispatched to inspect all ships, identifying crew members who were ill or suspected of being potential sources of an epidemic; these individuals were disembarked. This was the case for Admiral Juan Martínez de Recalde, who suffered from fevers that severely weakened his health[40] – a condition that would take its toll during the arduous return journey. The Armada finally set sail on 30 May, but with northern winds and unexpected downpours that slowed its progress, particularly for the heavy hulk ships. The constant storms culminated on 18 June when Medina Sidonia ordered the fleet to seek refuge in La Coruña – a necessary but unplanned stop. Under such conditions, managing an Armada with such a heterogeneous composition of vessels was nearly impossible. On 19 June, an unseasonal hurricane scattered over 100 ships along the Galician and Cantabrian coasts. For instance, the galleon *San Luis* was blown as far as La Rochelle in France.[41] Meteorologist Linés Escardó has conducted extensive studies on the weather faced by the Armada and identified three particularly violent phases of the storm: on the afternoon of 19 June, the morning of 23 June and throughout 1 July These storms caused significant damage to numerous vessels. It took three weeks to regroup all ships and carry out repairs. Remarkably, despite the continuous tempests, no ship was lost to shipwreck. Once repairs were completed and with favourable south-westerly winds, the Armada finally set course for England on Friday, 22 July. Upon departing La Coruña, the Armada was composed of the following fleets:

- The Portuguese Squadron, under the command of the Duke of Medina Sidonia, was militarily the most powerful. It consisted of ten war galleons – nine Portuguese-built: *San Martín* (flagship), *San Juan* (admiral's ship), *San Marcos*, *San Felipe*, *San Luis*, *San Mateo*, *Santiago el Menor*, *San Cristóbal*, and *San Bernardo* – and one Spanish-built galleon, *San Francisco*.

Additionally, it included two small and light zabras (*Augusta* and *Julia*) used for reconnaissance and courier missions.

- The Castilian Squadron, commanded by Diego Flores de Valdez, was another 'war fleet' composed of ten Spanish-built galleons: *San Cristóbal, Nuestra Señora de Begoña, San Medel y San Celedón, San Juan el Menor, Santiago el Mayor, Asunción San Felipe y Santiago, San Pedro Nuestra Señora del Barrio y Santa Ana*. It also included four naos (*Santa Catalina, Trinidad*, and two named *San Juan Bautista*) and two vessels (*Nuestra Señora del Socorro* and *San Antonio de Padua*)
- The Biscayan Squadron, under Admiral Juan Martínez de Recalde, served primarily as a transport fleet with ten naos: *Santa Ana, Gran Grin, Santa María de Montemayor, Santiago, María Juan, Magdalena, Manuela, Concepción Mayor o Zubelzu, Concepción de Joanes de Elcano,* and *San Juan*. It also included four vessels: *María de Miguel Suso, San Sebastián, Isabella,* and *María de Aguirre*.
- The Guipuzcoan Squadron, commanded by Miguel de Oquendo with ships built in Guipuzcoan shipyards, was another transport fleet consisting of nine naos: *Santa Ana, Santa María de la Rosa, San Salvador, Santiesteban, Santa Cruz, Santa María la Santa Bárbara, San Buenaventura,* and *María San Juan*. It also included one hulk (*Doncella*), three vessels (*San Bernabé, La Asunción*, and *La Magdalena*), and one pinnace (*Nuestra Señora de Guadalupe*).
- The Andalusian Squadron, commanded by Pedro de Valdés, combined warships and transport vessels. It was composed of the following ships: the *Nuestra Señora del Rosario, San Francisco, Duquesa Santa Ana, Concepción* and *San Juan de Gargoriu*; the galleon *San Juan Bautista*; the hulks *San Bartolomé, Santa Catalina, Santa María de Juncal*, and *Trinidad*; and the vessel *Espíritu Santo*.
- The Levant Squadron, commanded by Martín de Bertendona, was a transport fleet consisting of ten naos: *La Regazona, La Lavia, La Trinidad Valencera, Santa Trinidad de Scala, La Juliana, San Nicolás, La Rata Encoronada, San Juan de Sicilia* (or *Santo Joanne Battista*), *Anunciada,* and *Santa María de Visón*.
- The Galleass Squadron, under the command of Hugo de Moncada, was a war fleet composed of four galleasses: *San Lorenzo, Girona, Zuñiga,* and *Napolitana*.
- The Galley Squadron, commanded by Diego de Medrano, was also a war fleet consisting of four galleys: the *Capitana*, the *Bazana*, the *Diana*, and the *Princesa*.
- The Hulk Squadron, under Juan Gómez de Medina, comprised twenty-three hulks, mostly from northern Europe: *Gran Grifón, San Salvador, Castillo Negro, Casa de Paz Grande, Santiago, Barca de Hamburgo, San Pedro Mayor, Falcon Blanco Mayor, San Pedro Menor, Sansón, Barca de Ancique, David, San Andrés, El Gato, Ciervo Volante, Santa Bárbara, Casa de Paz Chica, Falcon Blanco Mediano, Esayas, San Gabriel, Paloma Blanca, Perro Marino,* and *Buenaventura*.

- The Vessels and Zabra Squadron, led by Diego Hurtado de Mendoza, was headed by the nao *Nuestra Señora del Pilar de Zaragoza*. It included two hulks (*La Caridad Inglesa* and the *San Andrés de Escocia*), thirteen vessels (*Santo Crucifijo, Concepción de Lastero, Espíritu Santo, Nuestra Señora de Fresneda, Concepción de Carasa, Concepción de Castro, Nuestra Señora de Guadalupe, San Francisco, Nuestra Señora de Begoña, Concepción de Capitillo, Nuestra Señora de Gracia, Nuestra Señora del Puerto*, and the *San Jerónimo*), and eight zabras (*Asunción, Concepción de Somarriba, Concepción de Valmaseda, Nuestra Señora de Castro, San Andrés, San Juan Carasa, Santa Catalina*, and *Trinidad*).

Problems soon reappeared. On the following day, 23 July, the galley *Diana* of the Galley Squadron broke her rudder stock and was forced to return to Spain, reaching Vivero (Galicia) without incident the next day. On Monday, 25 July, a pinnace under the command of Captain Rodríguez Tello departed for Dunkirk to inform Farnese of the fleet's arrival. Two days later, a new storm broke out, lasting two days and causing the dispersion of the Armada. By the morning of Thursday, 28 July, forty ships were missing from formation! Medina Sidonia ordered a group of vessels to search for the lost vessels. The fleet eventually regrouped 75 leagues (416km) from the Isles of Scilly. No shipwrecks were reported, but it was decided that the Galley Squadron should return to Spain. The *Bazana* was heavily damaged, and the remaining galleys were unable to continue sailing under such weather conditions. In fact, the *Diana* sank days later near Boulogne. The return of the galleys was a severe setback for the Armada, as it meant losing 200 cannons. On Friday, 29 July, the Andalusian Squadron and many of the delayed ships rejoined the formation. By that afternoon, the entire Armada had regrouped except for the nao *Santa Ana* (flagship of the Biscayan Squadron), which joined the following day. That afternoon, Lizard Point was sighted – the Armada had reached England.

The events of the Armada's voyage are detailed in two key accounts: Pedro Coco Calderón's *Sucesos de la armada desde la salida de La Coruña hasta la entrada en Santander, vistos desde la urca almiranta San Salvador*,[42] and Captain Alonso Vengas' *Sucesos de la Armada desde la salida de La Coruña hasta su regreso*.[43]

The Battles
Saturday, 30 July
At dawn, with little wind, Medina Sidonia ordered the sails to be furled to allow the delayed ships time to rejoin the Armada. He also convened a War Council to discuss the possibility of attacking the English

fleet – approximately ninety ships, about fifty of which were warships – anchored in Plymouth. Leyva and Recalde supported attacking the English, but the other captains preferred to proceed into the English Channel. Medina Sidonia, who had received orders from the King to avoid engagements and meet Parma as soon as possible, leaned toward entering the Channel, as strongly advised by Diego Flores de Valdés. With the arrival of the delayed ships, the Duke ordered the royal standard to be raised, displaying an image of Christ on the cross on one side and the Virgin on the other. Three cannon shots were fired, and prayers were said. The fleet then assumed a combat formation. The Armada entered the English Channel with 123 ships, following the withdrawal of the Galley Squadron and the vessel *Nuestra Señora de Gracia*. This decision by Medina Sidonia was later heavily criticised and considered one of the key factors in his mission's failure. It is worth noting that nearly a century later, in 1667, a Dutch fleet under Willem Joseph van Ghent and Michiel de Ruyter launched a decisive raid on England's fleet anchored at Chatham during the Medway Raid, significantly impacting the Second Anglo-Dutch War.

Sunday, 31 July

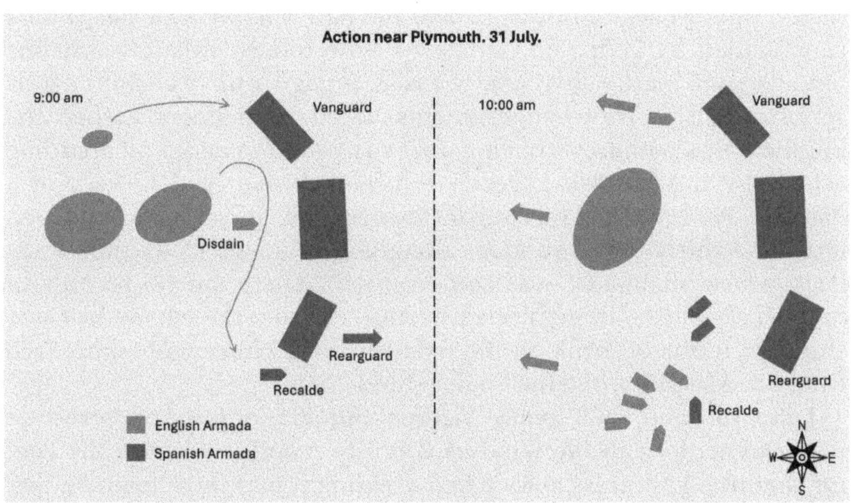

At dawn, the wind blew offshore, shifting throughout the day to the south-west. Part of the English fleet was sighted approximately four leagues (about six miles) leeward. The English manoeuvred so that, when the wind shifted, they positioned themselves windward with ninety ships – that is, to the west of the Armada. Closer to land, another eleven ships under Drake's command were spotted; he had departed slightly later. The Queen's fleet advanced in a disorderly manner, divided into two groups lacking any tactical formation. Meanwhile, the Spanish Armada maintained a compact formation in the classic pincer arrangement, which could resemble a crescent shape, with three major squadrons arranged from north to south: the Spanish vanguard, under Leyva, commanded the Levant Squadron, comprising about twenty ships; at the centre, Medina Sidonia's flagship *San Martín* closed the formation, flanked by transport and supply vessels; and in the south, the rearguard, led by Recalde, was composed of the Biscay Squadron with another twenty ships.

The opening act began with a provocation from the *Disdain*, a pinnace owned by Howard, which advanced to fire the first shot at the *Rata Santa María Encoronada* before retreating to its formation. Immediately afterward, a group of eight English ships advanced and began firing from a distance on Leyva's Spanish vanguard, though with little accuracy due to the range. Meanwhile, in the southern sector, Recalde's ship lagged behind – an old Spanish tactic of using a vessel as bait to provoke boarding and create a mêlée. Among the attackers was Howard's *Ark Royal*. Reinforcements arrived in the form of the galleon *San Mateo* under Diego de Pimentel; the galleon *San Juan el Menor*, under Diego Enríquez;[44] and the nao *Nuestra Señora de la Rosa*, commanded by Pedro de Garagarza, who fought valiantly. Medina Sidonia's *San Martín* also came to assist, engaging three enemy vessels simultaneously. These engagements lasted until noon,[45] when the English ships withdrew, having only succeeded in avoiding boarding actions by the Spanish galleons, which they managed to keep at a distance. As a result of this initial clash, Spanish losses included seven dead and thirty-one wounded, along with damage to Recalde's *San Juan*, whose mainmast was compromised. There are no records of English casualties, though Spanish sources claim the enemy lost two ships. In terms of artillery, the Spanish fired 720 cannon shots (120 from *San Martín*), while the English fired 2,000.

Later, around 2:00 p.m., Medina Sidonia ordered a series of manoeuvres to gain the windward position and to re-form the fleet for continued progress, once again dividing it into three main bodies: a vanguard, now led by Recalde's *San Juan* to allow for repairs; the centre, with transport vessels; and the rearguard, which held the majority of the warships and was closed by the *San Martín*. During these manoeuvres, two accidents occurred. The *Nuestra Señora del Rosario*, Pedro de Valdés's flagship of the Andalusian Squadron, collided with the *Catalina*, rendering the former ungovernable when her bowsprit was damaged and the foremast fell onto the mainmast.

Medina Sidonia ordered the galleass *Zuñiga* and several pinnaces to rescue the crew, but worsening weather conditions made the operation impossible, leaving the *Nuestra Señora del Rosario* adrift. Medina Sidonia could have ordered the fleet to halt to attempt a tow, but this would have disrupted the formation, as the vanguard was already moving ahead. On the advice of Diego Flores de Valdés, he decided to abandon the ship to its fate. It is worth noting that Pedro de Valdés was Diego Flores's cousin, and their personal relationship was notoriously poor. This decision shocked most of the squadron commanders, though they ultimately had no choice but to accept it. The second incident involved the *San Salvador*, flagship of the Guipúzcoa Squadron, which suffered a massive explosion that destroyed both decks and the sterncastle, killing at least 200 crew members. The cause of the accident remains unknown; it may have been accidental or the result of sabotage. A German gunner who had argued with his captain, Pedro de Pliego, was suspected. Medina Sidonia dispatched several pinnaces to recover survivors and valuable goods. The wounded were transferred to the *San Pedro Mayor*, a hulk serving as a hospital ship, while the remaining crew were distributed among various pinnaces. That same day, a pinnace was sent to Flanders to inform the Duke of Parma of the Armada's arrival. On the English side, Howard – unaware of the Armada's accidents – began pursuit at nightfall. He ordered Drake to lead the fleet, guiding them with a lantern. Of this initial encounter, Howard would later write: 'We did not dare to approach, for their fleet is eminently stronger.'

Monday, 1 August

In the early hours, the Spanish rearguard formed a compact nucleus of forty vessels under the command of Recalde. This group included Martín de Bertendona's Levant Squadron, the three galleasses (*San Lorenzo*, *Girona* and *Napolitana*) and four galleons from the Portuguese Squadron, as well as the *San Juan*, *San Mateo*, *San Luis*, *Santiago* and the *Florencia*. At the same time, the English fleet became dispersed. Drake's lantern, which served as a navigational reference, vanished from the view of the remaining ships, which were either lost, scattered, or simply delayed. Drake – who was, above all, a privateer rather than a soldier – upon seeing that the *Nuestra Señora del Rosario* was adrift, could not resist the opportunity to seize a valuable prize. Valdés, likely enraged with Medina Sidonia, chose to play his own hand and made the decision he deemed most favourable for himself and his crew: to surrender without resistance. He could have chosen to resist the *Revenge*, disabled his cannons, thrown the powder and even the money overboard. Instead, he negotiated advantageous terms of surrender with Drake – terms which the Englishman fulfilled scrupulously. The soldiers and crew were sent to various prisons in London, where they were released two years later. Pedro de Valdés, Vicente Álvarez (the captain and owner of the ship), and the other notable figures were held at Richard Drake's country estate, where they received excellent treatment until their release in 1593, after the payment of a substantial ransom. Furthermore, Sir Francis seized the 50,000 ducats the galleon was transporting – equivalent today to approximately £3,360,000.[46] However, Drake officially declared only 20,000 ducats (£1,344,538), leaving the fate of the remainder a mystery – though one easily speculated.[47] Not a bad way to start a Monday morning. Nevertheless, Drake's actions drew criticism from his peers. For instance, Frobisher wrote:

> Drake took Pedro, for after he had seen her in the evening, that she had spent her mast, then like a coward, he kept by her all night, because he would have the spoil. He thinkers to loosen us for our shares of 15 thousand ducats, but he will have our shares, or I will make him spend the best blood in his belly.[48]

At sunrise, Howard's *Ark Royal, Triumph, Bear* and *Marie Rose* realised they were within cannon range of the Armada. Moreover, a lack of wind exposed them to the Spanish ships. Medina Sidonia, whose sails were similarly affected by the calm, ordered the three galleasses in his rearguard to approach any of the isolated English vessels to engage them. However, the galleasses moved slowly, and the wind soon picked up again, allowing the English ships to adjust their sails and rejoin the vanguard. Historian García-Torralba[49] laments the absence of galleys, as had been desired by the Marquis of Santa Cruz, arguing that they could have attacked the dispersed English ships. Nevertheless, it must also be acknowledged that both the galleasses and galleys were poorly equipped with artillery, although they were excellent in boarding actions. In any case, had the twenty galleys demanded by Santa Cruz accompanied the Armada, there is no reason to believe they would not have been forced to retreat before reaching the English Channel – just as happened to the four that ultimately did participate. Meanwhile, Howard spent the entire day regrouping his scattered ships. By mid-morning, the *San Salvador* was boarded by English vessels. The *Golden Hind* towed her to the port of Weymouth, where the twenty-nine surviving crew members were disembarked. The foreign sailors were released, while the Spaniards were imprisoned. The ship sank in November, while *en route* to Portsmouth for repairs.[50]

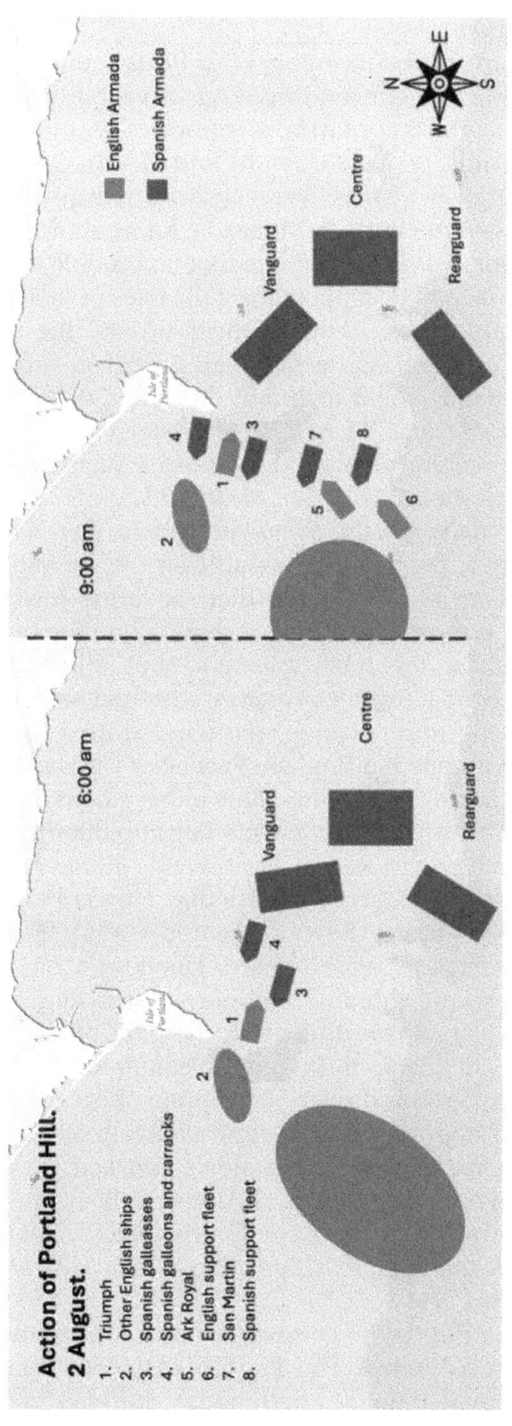

Action of Portland Hill.
2 August.
1. Triumph
2. Other English ships
3. Spanish galleasses
4. Spanish galleons and carracks
5. Ark Royal
6. English support fleet
7. San Martin
8. Spanish support fleet
English Armada
Spanish Armada
6:00 am
9:00 am
Vanguard
Centre
Rearguard
Vanguard
Centre
Rearguard
N
E
S
W

Tuesday, 2 August

In the early hours of the morning, both fleets found themselves near Portland Bill, in calm sea conditions, clear visibility, and an absence of wind. Aboard the San Martín, a council of war was held with the presence of Recalde, Leyva, Oquendo, and Medina Sidonia. Observing that several English vessels were separated from the main group, orders were given to Hugo de Moncada, admiral of the galleasses, to engage these isolated ships. Anticipating favourable winds at dawn, a group of galleons was to support the galleasses in their attack. Shortly before daybreak, in execution of these orders, the four galleasses advanced and attacked the most isolated English ships located near the Isle of Portland. These included Frobisher's *Triumph*, as well as the *Golden Lion of London, Mary Rose, Merchant Royal, Margaret and John* and *Centurion*, among others. At dawn, the wind rose in favour of the Spanish, and the galleons *San Medel* and *San Francisco de Florencia*, along with the naos *Nuestra Señora de Begoña, San Juan Bautista* and *Trinidad Valencera*, launched an assault on the *Triumph*. However, a strong counter-current hampered their advance, forcing Moncada's *San Lorenzo* to face several English ships alone for over an hour. It is likely that during this clash, a cannon shot killed Captain William Coxe of the *Delight*, a former privateer who had seen little fortune in the Caribbean. By mid-morning, the wind shifted, and the weather gauge came to favour the English. Frobisher's situation had become desperate, but capricious Aeolus came to his aid just as he was on the verge of being boarded. The Queen's flagship, along with fifty other vessels – including Sir Robert Southwell's *Elizabeth Jonas*, George Fenner's *Galleon of Leicester*, Lord Thomas Howard's *Golden Lion*, Sir John Hawkins's *Victory*, Edward Fenton's *Mary Rose*, Sir George Beeston's *Dreadnought*, and Richard Hawkins's *Swallow* – rushed to Frobisher's rescue. Medina Sidonia, recognising the manoeuvre, positioned his *San Martín* directly in the path of the English flotilla in an attempt to block their reinforcement of the *Triumph*. The Spanish flagship endured over 500 cannon shots over the course of more than two hours, of which approximately 50 struck the vessel. Soon, reinforcements arrived: Recalde's *San Juan*, Oquendo's *Santa Ana*, Agustín Mexía's *San Luis*, and Alonso Téllez Girón's *San Marcos*. With the wind now favouring the English, the *Triumph* managed to disengage from the Spanish ships, after which Howard ordered the withdrawal of his fleet.

After nearly a full day of combat, Spanish losses amounted to fifty dead and sixty wounded. The English casualties remain unknown. In terms of naval damage, both sides suffered relatively minor impairments, with no ships lost. The Spanish fleet continued its eastward course unimpeded. That very same day in Bruges, the Duke of Parma received Captain Rodrigo Tello de Guzmán, dispatched by Medina Sidonia, who informed the Duke that the Armada had already reached 48 degrees north latitude – in other words, it was nearing. Farnese immediately ordered his troops to embark at Nieuwpoort and Dunkirk.[51]

Wednesday, 3 August

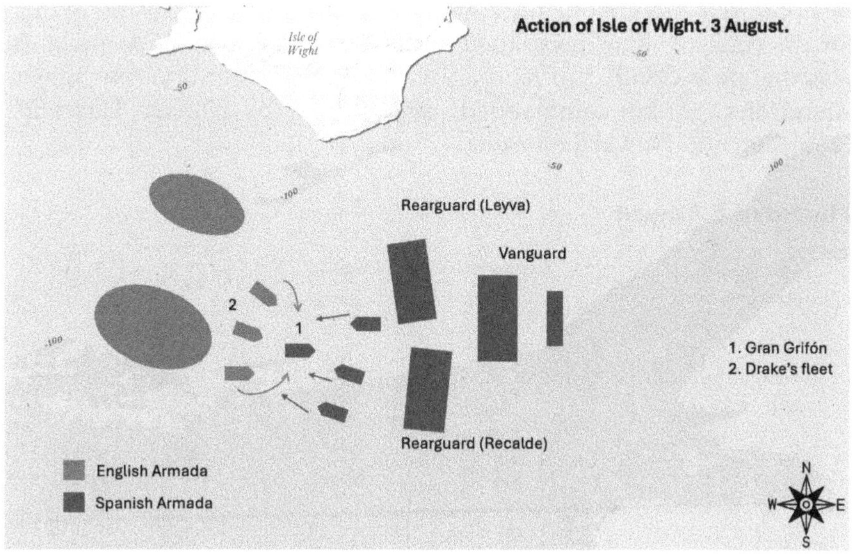

At dawn, the Armada continued its eastward advance toward the Isle of Wight. The formation was organised with two large squadrons at the vanguard, while two additional fleets brought up the rear. These were commanded by Leyva on the northern flank – with the squadrons of Oquendo and Bertendona – and by Recalde on the southern flank. During the morning, the hulk *Gran Grifón*, captained by Juan Gómez de Medina, fell behind the main group. Drake's flotilla seized the opportunity to descend upon the heavy and slow-moving Spanish vessel. The crew of the *Gran Grifón* endured an agonising hour during which it appeared they might be boarded. However, they managed to hold their attackers at bay through sustained musket and arquebus fire, supported by deadly short-range artillery. Eventually, the galleasses *San Lorenzo* under Moncada and the *Zúñiga*, as well as Recalde's galleon *San Juan*, arrived in support of the embattled hulk. A fierce artillery exchange ensued, during which, as was customary, Recalde placed his galleon in the most perilous position. Approximately an hour later, additional galleons and naos joined the fray, prompting Drake to conclude that the most prudent course of action was to withdraw to the protection of the Royal Navy's main force. In that engagement, the Spanish Armada reported seventy dead and sixty wounded. English casualties remain unknown.

By mid-afternoon, Howard convened a council of war with his admirals. If the aim was to pick off Medina Sidonia's ships one by one, it was clear that such a strategy was proving ineffective. The disorganised and chaotic formation of the English fleet was no match for the cohesive and disciplined structure of the Spanish Armada. It was therefore decided to reorganise the English fleet into four major squadrons, to be commanded respectively by Drake, Hawkins, Frobisher, and Howard himself.

Thursday, 4 August

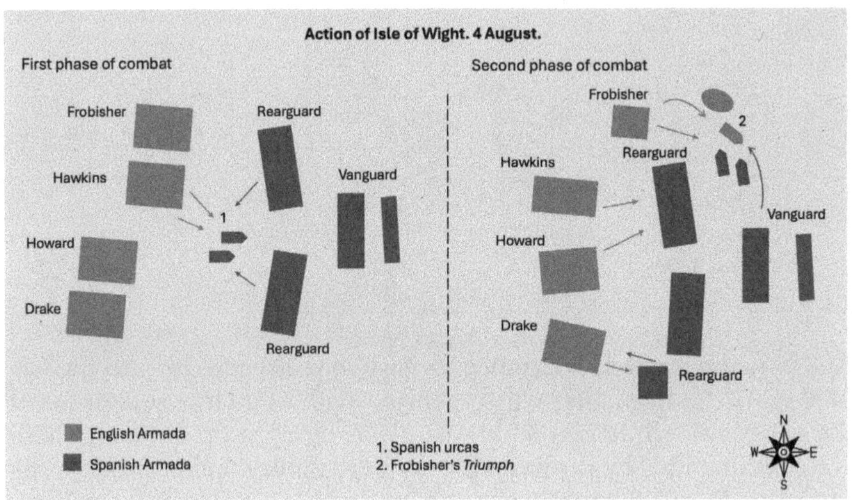

The day broke with scarcely any wind, causing several ships of the Armada to fall behind the main formation. Coco Calderón recounts that the heavy and slow-moving hulks *Santa Ana* and *Doncella* were left in a vulnerable position – an opportunity that was swiftly seized by the *Ark Royal* and five other warships of her squadron, which launched an attack on them. Hawkins, towed by smaller boats using oars, also sought to join in the harassment of the hulks. The first vessels to respond were the galleasses, which were propelled by oars, along with the nao *Rata Encoronada*, which was being towed by one of the galleasses. Suddenly, the wind began to blow. The

San Martín, along with several other galleons, joined the group advancing to assist the beleaguered hulks. For approximately two hours, the flagships of both fleets exchanged cannon fire, though without decisive outcome. In this skirmish, particular distinction was shown by the galleass *Napolitana*, under the command of Perucchio Morán. Meanwhile, to the north, Frobisher – taking advantage of the favourable wind – boldly outflanked the Spanish formation, reaching the level of the Spanish vanguard, where the *Gran Grin* of Pedro de Mendoza, the *San Juan de Sicilia* of Diego Téllez Enríquez, and the *San Cristóbal* of Gregorio de las Alas were stationed. The *Triumph* once again found herself in an extremely perilous position. Howard dispatched up to eleven rowing barges to extract the *Triumph* from danger. However, just as this operation had begun, a sudden shift in the wind allowed Frobisher to reunite with his flotilla – once again affirming that the Yorkshire seaman seemed to be Aeolus's favoured. Throughout the engagement, approximately 3,000 cannon shots were exchanged. Spanish casualties amounted to fifty dead and seventy wounded. The English losses remain unknown.

Friday, 5 August
In calm conditions and without wind, both fleets advanced slowly throughout the day without engaging each other. Medina Sidonia dispatched the pilot Domingo Ochoa with a new letter addressed to the Duke of Parma, requesting that he send between forty and fifty *flyboats*. The Duke planned to use these vessels for their speed and manoeuvrability to grapple onto English ships and immobilise them, allowing the Spanish galleons to approach and board.[52] Meanwhile, the English fleet, which had steadily incorporated numerous vessels from various southern English ports over the course of the week, now numbered 160 ships.

Saturday, 6 August
The two fleets sailed in proximity, maintaining both sailing and battle order, but no engagement occurred. Around 10 a.m., the Armada sighted the French coast near Boulogne, reaching the vicinity of Calais six hours later – approximately seven leagues from Dunkirk – where the Prince of Parma was supposedly awaiting them. The Spanish pilots advised against proceeding beyond that point due to the strong currents in the area. Medina Sidonia sent a message to the governor of Calais to inform him that the fleet had arrived without any hostile intent toward the French coast, to which the governor responded that the Armada was welcome. On the same day, Howard received reinforcements in

the form of thirty-six ships from Seymour's squadron, which had been stationed at the eastern end of the Channel.

Sunday, 7 August

At dawn, Medina Sidonia received Captain Rodríguez Tello, who carried a letter from the Duke of Parma stating that he was still in Bruges – 40 miles from Dunkirk – and would not be ready for another five days. An eternity. The passivity of Alessandro Farnese is striking, given that by 5 August the Armada was only 80 miles from Calais, and yet on the 7th he still had not prepared his troops for embarkation. Medina Sidonia urgently dispatched Jorge Manrique, the Armada's inspector general, to present the fleet's immediate needs to Parma and to urge him to act with the utmost haste. In a tone of reproach and desperation, he reminded Parma that he had been writing to him almost daily without receiving any reply. Furthermore, he reiterated his request for forty to fifty *flyboats* to engage the English fleet and thereby buy time for the two Spanish forces to unite.[53] Meanwhile, in Flushing, a fleet of thirty-two ships under the command of Justin of Nassau was monitoring Parma's movements. At the English anchorage, Howard convened a council of war to implement a plan (likely suggested by Sir William Winter): to launch fireships against the Spanish Armada. This was by no means a novel tactic. In fact, fireships had been used since Ancient Greece. It is probable that Winter, along with many of the admirals present at the council, was familiar with *Inventions or Devises*, published in 1576 by William Bourne, which detailed the use of such methods.[54] In any case, the Spanish admirals were fully aware of this possibility. One must recall that the Flemish rebels had used fireships against them during the Siege of Antwerp (1584–5). Therefore, that night, Medina Sidonia ordered the vessel and *zabra* squadrons, together with the longboats, to position themselves between the English and Spanish anchorages in order to guard against such an attack. That same night, Howard assigned Henry Palmer to sail in a pinnace to Dover to select the oldest vessels he could find. However, panic spread among the English admirals. Throughout their pursuit, they had been unable to break the Armada's compact formation. The few Spanish ships that had been lost were due to accidents rather than combat. Moreover, they had no reliable intelligence regarding the actual condition of Parma's army. Terrified by the prospect of the two Spanish forces successfully linking up, they resorted to a desperate solution: to select the fireships from within Howard's own fleet. The chosen vessels were the *Bark Talbot*, the *Thomas Drake, Hope Hawkins* of Plymouth, *Bark Bond, Cure's Chip, Bear Yonge, Elizabeth of Lowestoft*, and one smaller, unnamed ship.

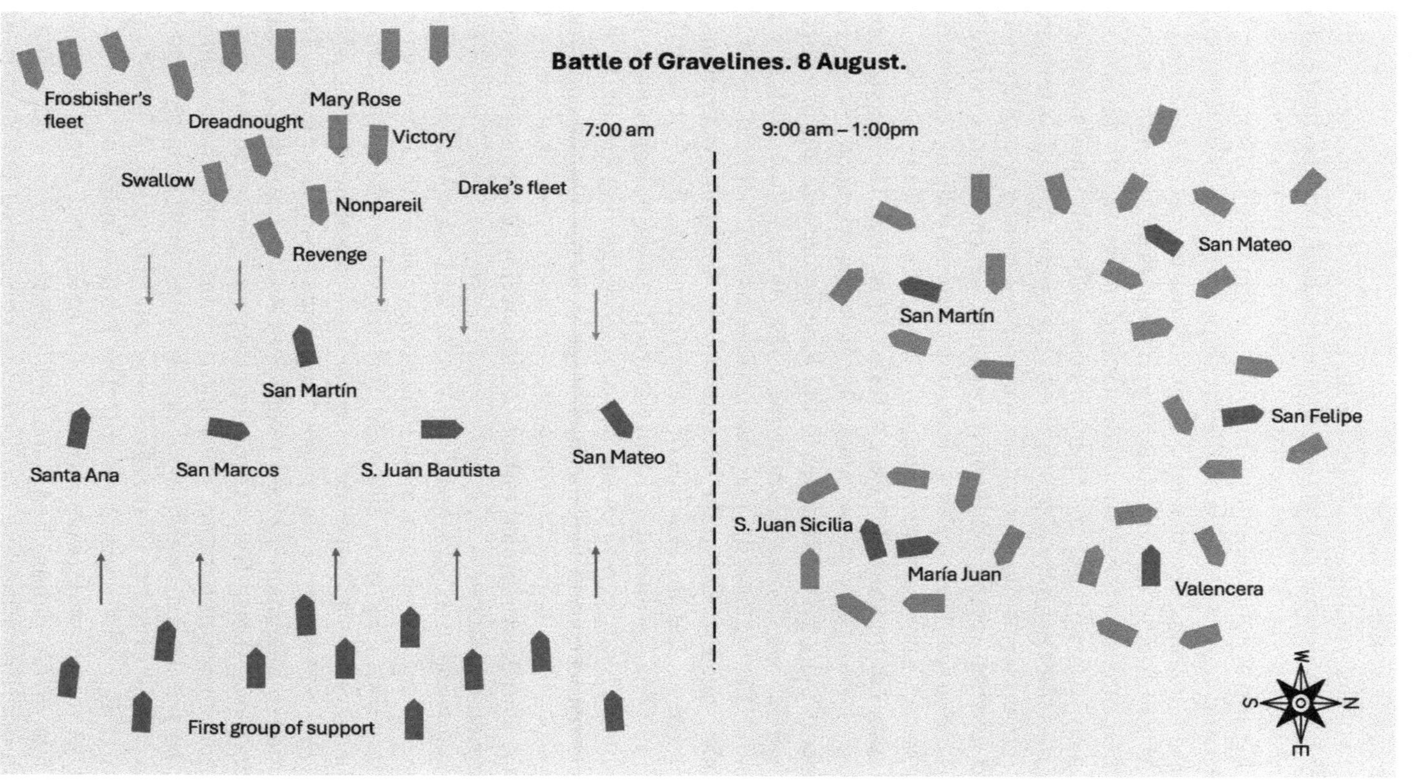
Battle of Gravelines. 8 August.
7:00 am
9:00 am – 1:00pm
Frosbisher's fleet
Dreadnought
Mary Rose
Victory
Swallow
Nonpareil
Drake's fleet
Revenge
San Martín
Santa Ana
San Marcos
S. Juan Bautista
San Mateo
First group of support
San Martín
San Mateo
San Felipe
S. Juan Sicilia
María Juan
Valencera
N
S
E
W

Monday, 8 August

At dawn, the fireships – smeared with pitch and tar and their cannons loaded with double charges of shot – reached the proximity of the Spanish Armada. The defensive line composed of vessels and *zabras* sprang into action, attempting to intercept the vessels using grappling hooks and harpoons. Two of the fireships were successfully stopped, but the remaining six managed to breach the barrier. As planned, the *San Martín* fired a cannon shot to signal the fleet to weigh anchor or cut their cables, as the situation demanded. Although the fireships did not succeed in striking any Spanish vessels directly, they nonetheless achieved their primary objective: breaking up the Armada's formation. The remainder of the disruption was carried out by a south-westerly wind and the region's powerful currents, which began to drive the ships out to sea. The resulting confusion caused a collision between the galleass *San Lorenzo*, under Hugo de Moncada, and the nao *San Juan de Sicilia*, commanded by Diego Téllez Enríquez. The impact broke the rudder of the *galleass*, rendering her ungovernable. Some crew members managed to transfer to the *Rata Encoronada*, but the current carried the *San Lorenzo* toward the coast, where she eventually ran aground. Listing heavily, she was unable to bring her artillery to bear for defensive fire. Given the circumstances – nighttime operations, no night vision, and the absence of radio communication – it is striking that only one significant incident occurred among a fleet of more than 120 ships. This undermines any hypothesis suggesting that the Spanish sailors panicked and fled the anchorage in terror. On the contrary, it suggests a high degree of professionalism and seamanship among the crews, who demonstrated notable manoeuvring ability even in such a precarious situation.

Meanwhile, the galleons *San Martín* and *San Marcos* managed to regain their positions at the anchorage, as Medina Sidonia had ordered. But they were alone. At this critical juncture, the Armada was engaged in a desperate struggle against nature itself – against the wind and currents that pushed the ships further out to sea. Three additional galleons remained relatively close to the ships of Medina Sidonia and Alonso Téllez Girón, but they were forced to tack upwind to rejoin their companions. From their vantage point, they could now see, to windward, some 130 English ships approaching – among them the flagships of Drake, Frobisher, and Hawkins. A catastrophe loomed. But where was Howard? The Lord Admiral of the English fleet, at the most decisive moment of a battle that might determine the fate of his nation, had vanished. Specifically, he had sailed off to plunder the stranded *San Lorenzo*, accompanied by some twenty-five ships from his squadron. The galleass, now listing heavily, had one side of her artillery pointing skyward and the other buried in the sand. Against Howard's fleet, the defenders could only rely on muskets and arquebuses. Approximately 150 sailors and galley slaves abandoned ship and made their way to the castle at Calais, where they were received by the governor, who would later hand them over to the Duke of Parma.[55] The English began to fire their cannons, and some of their crews disembarked to storm the vessel. A fierce firefight ensued, resulting in numerous casualties on both sides. The resistance continued until a single shot struck and killed Hugo de Moncada,[56] an event that broke the morale of his men, who then abandoned the vessel. There is little certainty regarding what occurred afterward. It is possible that a portion of the defenders reached the castle at Calais, but the majority – an indeterminate number – were captured by English forces. Some of the Spanish sailors and soldiers were executed, while forty were sent to English prisons.

While all this was taking place, cannon fire was already thundering across the sea. Between seven and eight in the morning, several ships had managed to form a defensive line intended to contain the fleets of Drake, Hawkins, and Frobisher. At first, in addition to the *San Martín* and *San Marcos*, the galleons *San Juan*, *San Mateo*, *San Luis*, *San Felipe*, *Florencia*, *Santiago* and *San Cristóbal*, along with the naos *Santa Ana*, *Trinidad Valencera* and *Ragazona*, succeeded in joining the formation. The first clash occurred between Drake's *Revenge* – followed astern by Fenner's *Nonpareil* and other ships of Drake's squadron – and the *San Martín*. Both vessels exchanged a tremendous broadside at very close range. According to Spanish sources, the *Revenge* was riddled by musket and cannon fire, receiving an estimated forty cannon hits. Following this brutal exchange, Drake withdrew toward the north-east and is not mentioned again in the sources for the remainder of the engagement. Once again, we see Drake's inability to abandon his so-called 'Caribbean' mindset: a reluctance to sustain combat on equal terms with an enemy. When the situation became complicated, he chose to flee and preserve his life – a conduct that, at Gravelines, even Frobisher criticised in his report to Lord Burghley, in which he remarked:

> Sir Francis reported that no man hath done any good service but he, but he shall well understand that others hath done as good service as he, and better too. He came bringing up at the first, indede, and gave them his prose and his broad side, and then kept his low and was glad that he was gone again, like a cowardly knave or traitor – I rest doubtful, but the one I will swear.

Yet Frobisher's and Hawkins's ships, along with the stragglers from Drake's squadron, continued to press against the Spanish line. With the wind at their backs, the English vessels began to encircle the Spanish galleons – including the *San Juan*, *San Felipe*, *San Mateo*, and *San Martín* – which were forced to fight under clear numerical disadvantage for several hours, at times outnumbered ten to one. Around ten o'clock in the morning, the fleets of Seymour and Howard joined the engagement. However, additional Spanish warships had also arrived – by that hour numbering sixteen, increasing to thirty as the morning progressed. Some sources claim there were as many as forty, though in any case, the English retained a numerical superiority of approximately three to one. Thanks to a north-westerly wind that blew throughout those hours, the Armada was able to reconstitute its once-impenetrable formation. The fighting was chaotic. From the accounts of Venegas and Coco Calderón, we know that the *San Martín* exchanged cannon fire with John Hawkins's *Victory*, Edward Fenton's *Mary Rose*, George Beeston's *Dreadnought*, Richard Beeston's *Swallow* and several other vessels. The *San Martín* was encircled – at times within arquebus range – but never boarded. Soon, a third English column appeared, led by Henry Seymour's squadron, with Seymour aboard the *Rainbow*, accompanied, among others, by William Winter's *Vanguard*. They targeted the starboard wing of the Spanish rearguard,[57] where the *San Cristóbal* of Gregorio de las Alas, flagship of the Castilian squadron, was likely positioned, alongside Bertendona's *Ragazona*, Diego Enríquez's *San Juan*, and Diego Téllez Enríquez's *San Juan de Sicilia*, inflicting severe punishment. Not long after, Howard's squadron appeared, led by the *Ark Royal*. Fortunately for the Spanish, more than two hours had already passed since the start of the engagement. However, many of the Armada's ships were being surrounded and relentlessly bombarded by the English. The *San Felipe* was caught in a crossfire by the *Elizabeth Bonaventure*, *Rainbow*, *Vanguard*, and others. The *San Mateo*, the *Valencera*, the *Santa María de Begoña*, and the *San Juan de Sicilia* turned to assist it. But in doing so, the *Begoña* became encircled by another group of English vessels.

It was then that the *San Martín*, under Medina Sidonia, and *La Rata*, commanded by Alonso de Leyva, appeared – almost miraculously – to extricate it from danger.[58] They then moved to rescue the *Santiago*, which was under attack by no fewer than seventeen English ships. The battle raged intensely until noon. Thereafter, it began to subside. Yet a new shift in the wind, this time from the south-east, began to push the Spanish ships not currently engaged toward the treacherous sandbanks along the Franco-Flemish coast between Calais and Nieuwpoort.

These were skilfully avoided thanks to the superior seamanship of the Spanish crews, once again demonstrating their navigational expertise. By around four in the afternoon, the wind began to intensify. But it was not until six o'clock that Howard – aware that a significant portion of his fleet was running low on powder and shot, and that he could do little more damage to the Spanish ships than had already been inflicted – ordered the withdrawal of his fleet.

As a result of the battle, the Armada lost three ships. The *María Juan*, a nao under the command of Pedro Sanz de Ugarte and belonging to the Biscayan Squadron, had suffered severe damage, having lost both her rudder and mizzenmast. The survivors signalled for assistance, and between fifty and sixty men were transferred to the *San Juan el Menor*, *San Juan de Sicilia*, and the *Magdalena*. As night fell, the vessel sank, resulting in the drowning of approximately 250 men, including her captain, Pedro Sanz de Ugarte.[59] Additionally, two galleons from the powerful Portuguese Squadron sustained irreparable damage and were ultimately lost. The first was the *San Mateo*, which had spent the entire day engaged in intermittent combat with thirteen English ships under Seymour and Winter, none of which dared to board her.[60] Observing the ship's condition, Medina Sidonia offered its captain, Diego de Pimentel, the opportunity to transfer his crew to other vessels. Pimentel declined, expressing confidence that he could carry out the necessary repairs. A small number of men did abandon the ship.[61] During the night, the *San Mateo* was driven toward Ostend. At dawn, a fleet of five Anglo-Dutch ships under Pieter van der Does[62] attacked her until the Portuguese galleon ran out of gunpowder.[63] It was then boarded, and a fierce fight ensued in which approximately forty Spaniards were killed. The survivors – around 300 men, including Diego de Pimentel – were taken prisoner. They would return to Spain months later, after a ransom had been paid. The *San Mateo* was beyond repair and eventually sank. The second Portuguese galleon lost was the *San Felipe*, commanded by Francisco de Toledo, brother of the Count of Orgaz. According to Coco Calderón, the hulk *Doncella* approached, enabling the evacuation of a large portion of the crew. However, Francisco de Toledo observed that the hulk was also in poor condition. In response, Juan Posa, an officer of the Tercios, reportedly declared – again according to Coco Calderón – that 'if I am to drown on a hulk, I would rather drown on a galleon', and he remained aboard the *San Felipe* with his captain and about thirty other men. Eventually, the ship ran aground on a beach between Nieuwpoort and Ostend. The remaining crew abandoned the vessel; some reached the Duke of Parma's camp, while others were captured and taken alongside the survivors of the *San Mateo*. It must be noted, of course, that all galleons and naos that had taken part in the battle suffered varying degrees of damage. For instance, the *San Martín* was struck by 108 cannon shots, and the *Santa María de la Rosa* sustained nearly as many. Many hulls and masts were severely damaged. Spanish casualties were estimated at 600 dead and 800 wounded. In percentage terms, the Spanish had lost 2.34 per cent of their ships in combat over the course of the week (a figure that increases slightly when including those lost to accidents). As for English losses, virtually nothing is known. The Queen had imposed a strict veil of silence, under threat of imprisonment or even death. However, a dispatch intercepted on 26 August provided some insight into the condition of the returning English fleet: '[to London] 28 ships very badly damaged, and to Pechelingas [Flushing] 32 in even worse condition and with sparse crews, many of whom had died, including their chief pilot; and that the Queen had issued a proclamation forbidding anyone in her realm from speaking of the outcome of the Armada, or from allowing ships to depart from her ports ...' In any case, the Spanish were no longer in the English Channel. But was England truly safe?

And Parma? Where was the Duke of Parma? Alessandro Farnese was in Nieuwpoort (later moving to Dunkirk), coordinating the embarkation operation. At dawn, he received Jerónimo de Arceo, Medina Sidonia's personal secretary, who requested ships, gunpowder, and ammunition to assist in the battles.[64] At that moment, Parma realised the magnitude of the disaster and how disconnected from reality the King and his advisors were. He could not provide assistance – he needed it himself! Farnese had grown weary of requesting more time from the King to prepare for the invasion. Up until then, the ships he had managed to gather were entirely unsuitable for navigating the English Channel. Some were leaking, and others threatened to sink if faced with rough seas. Moreover, nearly all of them were flat-bottomed barges with flush decks that lacked masts, sails and cannons, requiring towing by Medina Sidonia's fleet – which itself could not approach the ports of Dunkirk and Nieuwpoort due to their shallow waters rather than English naval action. This explains Farnese's insistence on capturing the ports of Walcheren or Flushing;[65] without them, a rendezvous was impossible. Meanwhile, in the morning, obedient soldiers stood ready for embarkation. In Dunkirk, 6,000 soldiers from the Spanish Tercios were joined by Sir William Stanley's Irish Tercio (which included English and Scottish soldiers). In Nieuwpoort's port were Italian, German, Walloon, and Burgundian contingents[66] that, along with cavalry units, totalled approximately 10,000 men. The soldiers remained aboard their barges until evening when the Battle of Gravelines had already concluded, and winds were carrying the Armada toward the North Sea. By nightfall, orders were given to disembark. Some soldiers were discharged, but most remained in Parma's service. With them, it was hoped that the war against the Dutch rebels – already nearly won – could be concluded. However, fate had reserved a bitter end for the Duke, as we shall discover later.

The Blue Inferno: The Return

In the early hours of Tuesday, 9 August, a strong west-north-west wind arose, pushing the ships toward the island of Zeeland. The exhausted Spanish sailors were compelled to make a renewed effort – not only to avoid straying too far from the coast, but also to prevent getting dangerously close to it. At dawn, the wind subsided, affording the fleet a brief moment of respite. The rearguard group was composed of the *San Martín*, *San Juan de Recalde*, *San Juan Bautista*, *San Marcos*, *La Rata Encoronada* and the *Girona*. From their stern, they sighted Howard's fleet, though on this occasion they counted only 109 sails. Shortly thereafter, the English ships approached, prompting Medina Sidonia to order the Spanish vessels to head into the wind – forcing the enemy ships to maintain their distance. Simultaneously, he dispatched a vessel to the leeward ships (those further west), instructing them to sail as close to the north-west wind as possible, for they were dangerously near the shoals. The situation was so critical that, upon reaching a depth of only six and a half fathoms, the pilots nearly gave the Armada up for lost, fearing that a significant portion of the fleet might run aground – an outcome that would constitute a true disaster.[67] These were agonising moments. According to Coco Calderón, aboard the *San Martín*, some even advised Medina Sidonia to initiate peace negotiations with Howard. The Duke flatly refused, responding that 'he had faith in Our Lord and His Glorious Mother to bring him to a safe harbour; and should the Lord will otherwise, it would never be said that he acted contrary to the example of his ancestors'. With surrender ruled out but without a clear course of action, Medina Sidonia turned to Oquendo and asked, 'Señor Oquendo, what shall we do? For we are lost.' Oquendo replied that he should ask Diego Flores, but that he himself was prepared 'to fight and die like an honourable man. Let Your Excellency order me supplied with shot.' Miraculously, around 11:00 a.m., the wind shifted to west-south-west, enabling the Armada to head for the open sea and escape the shoals. For the time being, disaster had been averted. Later that afternoon, a war council was convened aboard the *San Martín*, attended by the principal admirals, ship captains, officers of the Tercios, and the fleet's pilots. They were tasked with deciding the Armada's next course of action. Any attempt to reestablish contact with Parma's fleet was ruled out. Recalde and Leyva advocated for retracing their steps and returning to Spain via the English Channel. However, the pilots maintained that, given the adverse winds and currents, a fleet of that size – navigating close-hauled (i.e., slowly and against the wind) – and with ships as unmanageable as the hulks under such conditions, the plan was not viable. Considering these factors, along with the shortage of ammunition and the extensive damage sustained by most vessels, Medina Sidonia decided that the fleet would return to Spain by sailing around the coasts of Scotland and Ireland, despite this route being approximately 750 leagues across perilous and largely unknown waters. A parallel war council was held aboard the *Ark Royal*. The English commanders, uncertain of the Armada's intentions, resolved to maintain pursuit. Meanwhile, Seymour's squadron remained stationed in the English Channel to monitor the movements of the Duke of Parma, who, for his part, had convened with his trusted officers and with Jorge Manrique, an envoy from Medina Sidonia, who was being made to understand that his *flybots* could not put to sea under current conditions. Their deployment would only be possible with favourable weather and in the absence of the enemy fleet. To attempt it under present circumstances would be tantamount to suicide.[68]

Near Ostend, though the exact date and circumstances remain unknown – most likely on 9 August – the vessel *San Antonio de Padua*, belonging to the Castile squadron, was lost. This small vessel did not participate in the combat, and it is therefore assumed that it ran aground on a beach, driven by wind and currents. According to a report from Jorge Manrique in Dunkirk, twenty-six crew members were murdered by Dutch soldiers,[69] while the remaining twenty-one were taken prisoner. It was also likely on 9 August that an incident involving several ships of the Armada took place. Despite Medina Sidonia's explicit orders to maintain a compact fleet formation, some ships disobeyed and broke ranks. The resulting breach of discipline so enraged the Commander-in-Chief of the Armada that he ordered the arrest of the responsible officers. These men were tried aboard the *Santa Ana*, and twenty captains were sentenced to death.[70] However, only one execution was carried out on that day: Cristóbal de Ávila, of the hulk *Santa Bárbara*, was hanged. The dire circumstances the Armada would soon endure prevented further executions from taking place. In the days that followed, monotony prevailed. Medina Sidonia remained fixated on preserving the fleet's tight formation, while Howard's fleet maintained close surveillance, occasionally approaching the Spanish rearguard. When challenged, however, the English ships would quickly fall back. This pattern persisted until 12 August, when the English fleet altered course and presented its stern to the Armada – signalling the end of the pursuit. At court, the Queen and her advisors vacillated between the uncertainty surrounding the Armada's actual condition and their own desire to begin demobilising the English fleet. In this context, Howard was ordered to return to port. Walsingham wrote to Lord Burghley: 'It is hard now to resolve what advice to give Her Majesty for disarming, until it shall be known what is become of the Spanish fleet'; and to the Lord Chancellor: 'I am sorry the Lord Admiral was forced to leave the prosecution of the enemy through the wants he sustained. Our half-doings doth breed dishonour and leaveth the disease uncured.'[71] The Queen's fear of invasion quickly gave way to her desire to uphold a strict fiscal policy. At that moment, the English fleet still comprised 119 ships at sea, with a total of 11,120 men.[72] But the celebration for having averted the invasion of England was short-lived for the sailors and crews. Howard grew increasingly desperate while waiting for Lord Burghley to send the funds needed to pay his men:

> There is a number of poor men of the coast towns – I mean the mariners – that cry out for money, and they know not where to be paid. I have given them my word and honour either the towns shall pay them or I will see them paid. If I had not done so, they would run away from Plymouth by thousands. I hope there will be care had of it. Sir, money had need to come down for our whole company.

To this, Walsingham warned: 'Sir, I do not see but of necessity there must be a magazine at Dover.'[73]

And the sailors believed him and waited to receive their wages. After all, Howard was an honourable gentleman with an impeccable reputation, not one of those adventurers with whom sailors usually embarked for the Caribbean in search of fortune. To make matters worse, an outbreak of typhus erupted aboard the *Elizabeth Jonas*, and it soon spread throughout the rest of the fleet. English sailors began to die by the dozens, then by the hundreds. Howard insisted: 'Sickness and mortality begins wonderfully to grow amongst us; and it is a most pitiful sight to see, here at Margate, how the men, having no place to receive them into here, die in the streets . . . It would grieve any man's heart to see them that have served so valiantly to die so miserably.'[74] Hawkins joined the Admiral-in-Chief's protests and asked Burghley whether he expected that 'by death, by discharging of sick men, and such like ... there may be spared something in the general pay'.[75] In this context, Elizabeth I delivered her famous speech at Tilbury – a brilliant piece of propaganda – in which she declared her willingness to die in defence of England. Yet this declaration came only after the danger had passed and as her army was being demobilised. Meanwhile, her sailors were dying – utterly neglected by the authorities – of hunger and disease. To alleviate the financial crisis, the Queen suggested attacking the Spanish Treasure Fleet, but her advisors responded that no ships were fit for the task.[76] Ironically, it may have been the typhus epidemic that saved the realm: dead men draw no pay. Meanwhile, Howard was using his personal fortune to pay his men and mitigate the effects of the illnesses ravaging the English ports. But beyond his charitable sentiment lay a practical concern:

> It were too pitiful to have men starve after such a service. I know her Majesty would not, for any good. Therefore, I had rather open the Queen's Majesty's purse something to relieve them, than they should be in that extremity; for we are to look to have more of these services; and if men should not be cared for better than to let them starve and die miserably, we should very hardly get men to serve.[77]

The war was not over. Abandoned by their Queen, it is estimated that some 8,000 men[78] died in the ports of southern England – men who had fought against Philip II's fleet. His own treatment of Spanish prisoners in England and Ireland, as we shall see, was markedly different.

However, the situation aboard the Spanish ships was far from favourable. Essential supplies for the crews' survival – water, food, firewood for cooking, and warm clothing – were becoming scarce. Hunger and disease had taken hold of the fleet. On 13 August, in a desperate measure to conserve fresh water, it was decided to throw overboard the mules intended to transport artillery upon landing in England, as well as the horses that many nobles had brought with them. This led to harrowing scenes in which distraught sailors cast the animals into the sea, and the terrified horses, in vain, attempted to return to the ships. In the following days, the weather remained reasonably favourable. However, by 18 August it was discovered that the *Rata Encoronada* and *San Juan de Sicilia* were no longer sailing with the fleet. On 21 August, Medina Sidonia dispatched Baltasar de Zúñiga[79] in a vessel to inform the royal court of the operation's failure and of the Armada's condition. Although the chosen vessel was particularly swift,

Portrait of Philip II of Spain by Sofonisba Anguissola (1573). (Museo del Prado. Public Domain)

Portrait commemorating the defeat of the Spanish Armada, depicted in the background. Elizabeth's hand rests on the globe, symbolising her international power. One of three known versions of the 'Armada Portrait'. (Public Domain).

Portrait of Álvaro de Bazán, Marquis of Santa Cruz, by Rafael Tegeo (1828). (Museo Naval de Madrid. Public Domain)

Landing of the Spanish Tercios on Terceira Island in 1583. (Fresco. Monastery of El Escorial. Public Domain)

Portrait of Alessandro Farnesse, Duke of Parma, by Otto van Veen dit Vaenius (1585). (Public Domain)

Francis Drake, by an unidentified painter, c. 1581. (National Portrait Gallery, London. Public Domain)

Portrait of the 7th Duke of Medina Sidonia, painted in 1612 by Francesco Giannetti. (Medina Sidonia Palace in Sanlúcar de Barrameda. Public Domain)

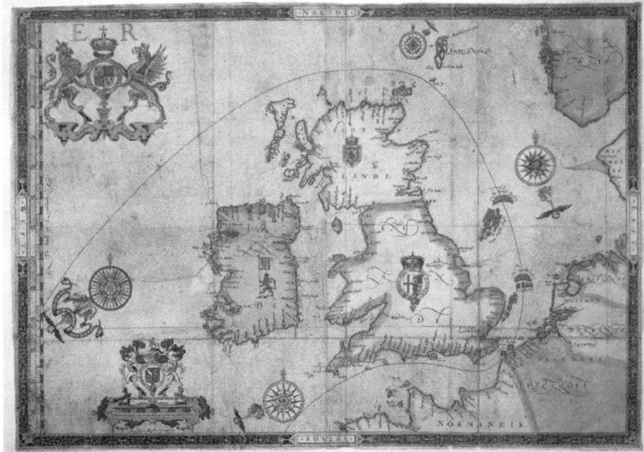

Following the Spanish Armada's defeat in 1588, Petruccio Ubaldini authored a brief account of the campaign, accompanied by a series of eleven charts by Robert Adams. These maps depicted the key naval engagements between the Armada and the English fleet, as well as the fleet's perilous retreat around Scotland and Ireland. The charts, engraved and published by Augustine Ryther in 1590, marked significant battle sites and shipwrecks. (Public Domain)

Admiral Pedro de Valdés surrendering his sword to Francis Drake aboard *Revenge* during the attack of the Spanish Armada, 1588. (Oil on canvas by John Seymour Lucas (1889). Public Domain)

One of the tapestry hangings of the House of Lords. It illustrates the different tactics employed by each fleet. While the Spanish formation is complex, the English formation is simple, merely pursuing the Spanish with little order. (John Pine, 1739. © Parliamentary Art Collection. Public Domain)

William I, Prince of Orange by Adriaen Thomasz. (Museo Thyssen-Bornemisza. Public Domain)

Elizabeth I and the Spanish Armada is an anonymous oil painting (121.3 × 284.5cm) once misattributed to Nicholas Hilliard and held by the Worshipful Society of Apothecaries of London. It presents a stylised montage of the battle of Gravelines, including beacons, Elizabeth's Tilbury speech, and the naval combat. (Public Domain)

Defeat of the Spanish Armada by Philip James de Loutherbourg (1789). (Public Domain)

Pennant of the Portuguese Galleon *San Mateo* of the Spanish Armada. With dimensions of 391 x 294.5cm (approximately 12ft 10in by 9ft 8in). (Museum De Lakenhal, Leiden, The Netherlands. Public Domain)

The Invincible Armada, 1588. Destruction of the Armada by the storm off the coast of Ireland, by Gartner de la Peña. (Museo de la Aduana en Málaga. Public Domain)

Sir Francis Walsingham by John De Critz the Elder (1589). (Public Domain)

Charles Howard, Lord Howard of Effingham (1536–1624), by Daniel Mytens (1620). (Public Domain)

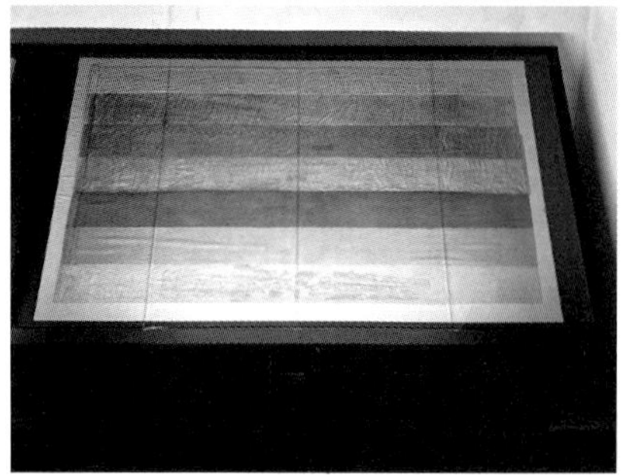

Known as the Flag of Sir Francis Drake, the only surviving sixteenthth-century naval flag is preserved in Sigüenza Cathedral. Captured by Captain Sancho Bravo de Arce during the failed English Counter-Armada expedition to Lisbon in 1589, it was later placed in his family's chapel. Rediscovered in 2011 by historian Luis Gorrochategui, it was restored and is now on display in the cathedral's museum. (© Catedral de Sigüenza/ ARTISPLENDORE)

María Pita Charging Against the English, by Arturo Fernández Cersa. (Patrimonio Artístico Municipal da Coruña. Public Domain.)

The Somerset House Conference representatives, 19 August 1604, by Juan Pantoja de la Cruz. (National Maritime Museum. Public Domain)

the journey was perilous and fraught with difficulty. So much so that it arrived in Santander three days after Medina Sidonia's flagship, the *San Martín*. From 24 August onward, weather conditions deteriorated rapidly. One violent storm followed another, and the Armada grew increasingly dispersed.

Upon reaching Ireland, disaster struck. Between 27 and 29 August, a group separated from the fleet, including the nao *Trinidad Valenzera* (Captain Alonso de Luzón), the hulk *Castillo Negro* (Captain Pedro Ferrat), *Gran Grifón* (Captain Juan Gómez de Medina), and *Barca de Hamburgo* (Captain Beltrán del Salto), the latter being the first to be wrecked. On 31 August, the hulk fired two cannon shots to signal that she was sinking. In response, the *Trinidad Valenzera*, *Gran Grifón*, and another hulk (likely the *Castillo Negro*) came to her aid. Crews were transferred: around 100 to the *Valenzera* (including Captain Beltrán del Salto), eighty to the *Gran Grifón*, and an undetermined but small number to the third hulk. Approximately ninety men remained aboard the *Barca de Hamburgo*, who were swallowed by the sea that night as they attempted to reach Malin Head, in northern Ireland. The *Trinidad Valenzera* eventually ran aground in Kinnagoe Bay[80] on 14 September. She remained afloat for several days, during which most of its 520 crew members were brought ashore safely. However, the ship ultimately sank, taking with it fifty men who had not yet disembarked, as well as some thirty Irish locals who had boarded the ship to loot it. The survivors made their way toward Alliagh Castle but were intercepted by a battalion of English soldiers and Irish mercenaries. Captain Alonso de Luzón negotiated a surrender in exchange for the lives of his men. As prisoners, they were led to Dublin. Along the way, the officers and noblemen – some fifty individuals – were separated from the rest of the crew, which numbered approximately 400 men. The following morning, around 250 of these men were brutally massacred by the English. A small group of survivors escaped and returned to Alliagh Castle, where they were sheltered by the Catholic bishop of the city, Conor O'Devany.[81] The bishop managed to arrange the transport of around 100 *Valenzera* crew members to Scotland, from where they were able to return to Spain. Another group, composed of the sick and wounded who could not travel, remained at the castle, though their fate is unknown. The officers and noblemen were sent to the notorious prison at Drogheda, following a gruelling march during which some of the captives perished. In prison, several died of disease or were executed. Ultimately, around thirty survivors were repatriated to Spain in 1590 after a ransom had been paid.

In early September, the galleon *San Juan* under Recalde also became separated from the main group. A vessel was dispatched in search of Medina Sidonia, but it returned without having located him. During those days, Recalde gradually assembled a group of vessels, eventually numbering nineteen sails. However, this formation would soon begin to lose ships. Meanwhile, the *San Juan* was forced to dock at an Irish port to carry out urgent repairs and to replenish her dwindling supplies of food and water. As for the galleass *Zúñiga*, she requested and was granted permission by Recalde to proceed independently. We shall return to this ship later. Another vessel that broke away was the hulk *San Pedro el Menor*, commanded by Captain Juan de Monsalve Solís.[82] Severely battered by the storms and abandoned by both her pilot and master in Calais,[83] she miraculously reached the French Breton port of Morbihan[84] on 20 September. There, it was aided by Bernardino de Mendoza, the Spanish ambassador in Paris, who arranged shelter and provisions for the crew. Eventually, the sailors, soldiers, and all salvageable materials were transported to La Coruña on 27 December, arriving four days later.[85] Three additional ships – *La Lavia, La Juliana,* and *Santa María de Visón* – also detached from Recalde's group. Aboard *La Lavia* was Francisco de Cuéllar, one of the ship captains previously condemned to death by Medina Sidonia, though the sentence was never carried out. A survivor of the ordeal, Cuéllar, would later recount the harrowing experiences endured by his fellow crew members.[86] According to Cuéllar, on 17 September the three ships anchored near the beach of Streedagh. The condition of *La Lavia* was disastrous, and it was agreed to transfer her crew to the other two vessels. However, during the operation, the three ships collided and sank, resulting in the deaths of approximately 1,000 men. Around 300 reached the shore, only to be met by a battalion of English and Irish soldiers who proceeded to massacre them. Only eighteen crew members managed to escape with their lives, finding refuge among the Catholic O'Neill and O'Donnell clans. Some Spaniards chose to remain and fight as mercenaries alongside these clans until 1596. Others, such as Francisco de Cuéllar, Manuel Orlando (captain of *La Lavia*) and Juan Bartoli (captain of *Santa María de Visón*), returned to Spain as soon as they were able. While this tragedy unfolded, a few miles to the south, at the mouth of the River Shannon, another group arrived: the nao *La Anunciada*, accompanied by the hulks *La Caridad* and *La Barca de Danzig*, and the *vessels La Concepción, Nuestra Señora de Begoña,* and *San Gerónimo.* Since *La Anunciada* had multiple irreparable leaks, it was decided – much like in the previous case – to redistribute her 266 crew members among the remaining four vessels. In this instance, the operation was completed without incident. The nao was burned on 21 September, and the rest of the flotilla successfully returned to Spain. That same day, *Santa María de la Rosa*, under the command of Martín de Villafranca, was shipwrecked near Dunmore Head. Only two men survived; they were captured and imprisoned in Galway, where one was executed and the other later released. The following day, Admiral Recalde's remaining, loosely cohesive group passed through the area and confirmed that there were no survivors from Villafranca's ship. Within this group was another nao in dire condition: the *San Juan Bautista* of Fernando Home, part of the Castilian squadron. The vessel had lost its mainmast and was so badly damaged that repairs were deemed impossible. Consequently, the ship's food, gunpowder, and crew were transferred to Recalde's *San Juan* galleon and to another Castilian nao also named *San Juan Bautista*. On 26 September, Fernando Home's ship was abandoned and set ablaze.

It was also on 17 September that *La Rata Encoronada* ran aground in Tullaghan Bay. On board was Alonso Martínez de Leyva, arguably the most important military commander of the Armada and the man King Philip II had designated as Medina Sidonia's successor should the Duke be unable to continue in command. Leyva's considerable prestige had attracted a significant number of nobles to this Genoese nao. With no alternative, Leyva ordered the evacuation of all personnel and subsequently commanded that the ship be burned. The Spanish forces fortified themselves in a nearby location, from which they sighted the hulk *Duquesa Santa Ana*, with which they managed to establish contact and arrange for evacuation. The details remain unclear, but thirteen members of the crew were captured and executed in Galway prison.[87] The hulk, now dangerously overloaded with more than 600 men aboard, ran aground a week later in Loughros More Bay. Once again, Leyva organised the defence of his men, this time among the ruins of a castle near Kiltoorish Lake. There, he learned that the galleass *Girona* was anchored some 15 miles away at Killybegs, undergoing repairs. The Spanish contingent made its way to Killybegs, and once the *Girona* was restored to seaworthiness, she took on board the survivors from *La Rata Encoronada* and the *Duquesa Santa Ana*, bringing the total number of crew to approximately 1,200 men. A further 200 men were left behind on land. Due to the extreme overloading of the vessel, Leyva judged it more prudent to attempt a return to Scotland rather than venture through the perilous Irish waters. However, on

24 October, during a violent storm, the *Girona* lost her rudder and, at approximately four o'clock in the morning, struck the rocks at Lacada Point (in what is now Northern Ireland). Only between three and nine men are believed to have survived. The sinking of the *Girona* constitutes the second deadliest maritime disaster in Spanish history in terms of loss of life.[88] Among the dead was Alonso Martínez de Leyva.

The fate of the remaining ships wrecked along the Irish coast shares a common denominator: shipwreck followed by the systematic execution of survivors. Such was the end met by the crew of the *zabra La Trinidad* on 15 September in Tralee Bay. That same day, off Valentia Island in southern Ireland, the vessel *Nuestra Señora del Socorro* disappeared at sea, with eleven men drowning and twenty-four others executed.[89] Five days later, the nao *Santisteban* of the Guipuzcoan Squadron wrecked near Doonberg, in County Clare. Not far from there, near Mutton Island, the Portuguese galleon *San Marcos*, one of the vessels that had seen the most combat in the English Channel, struck the reefs and sank with nearly her entire crew. Only four survived, of whom three were executed. Just one crew member ultimately managed to return to Spain. The *Gran Grifón*, a nao of the Biscayan Squadron, was wrecked on Clare Island. Of her 330 crew, only about 100 survived, but they were captured by the O'Malley clan, who immediately executed sixty of them, including the ship's captain, Pedro de Mendoza. The remaining survivors were sent to Galway prison, but only two would leave it alive and return to Spain. The hulk *Santa Bárbara* is known to have vanished in Ireland, though her fate remained unclear. However, the historian Chinchilla[90] discovered evidence that she may have run aground on 20 September near Carna. His research indicates that five prisoners from the ship were documented – three of whom were executed, and two repatriated to Spain – suggesting that the remaining forty-five crew members likely perished in the wreck. On 22 September, in Kilalla Bay, the hulk *Ciervo Volante* met her end: approximately sixty men drowned, eighty were executed, and sixty-two were imprisoned but later repatriated to Spain. On 25 September, the hulk *Falcón Blanco Mediano* wrecked on Freaghillaun South Island. Of the twenty survivors, all were imprisoned in Galway: only five managed to return to Spain, with the rest being executed. The following day, at Toorglass, the *San Nicolás Prodanelli*, a vessel of the Republic of Ragusa, ran aground. Of the approximately seventy crew who reached the shore, half were massacred by an English detachment, and the other half perished in Galway prison. Only two men – those who managed to evade capture – ultimately returned to Spain.

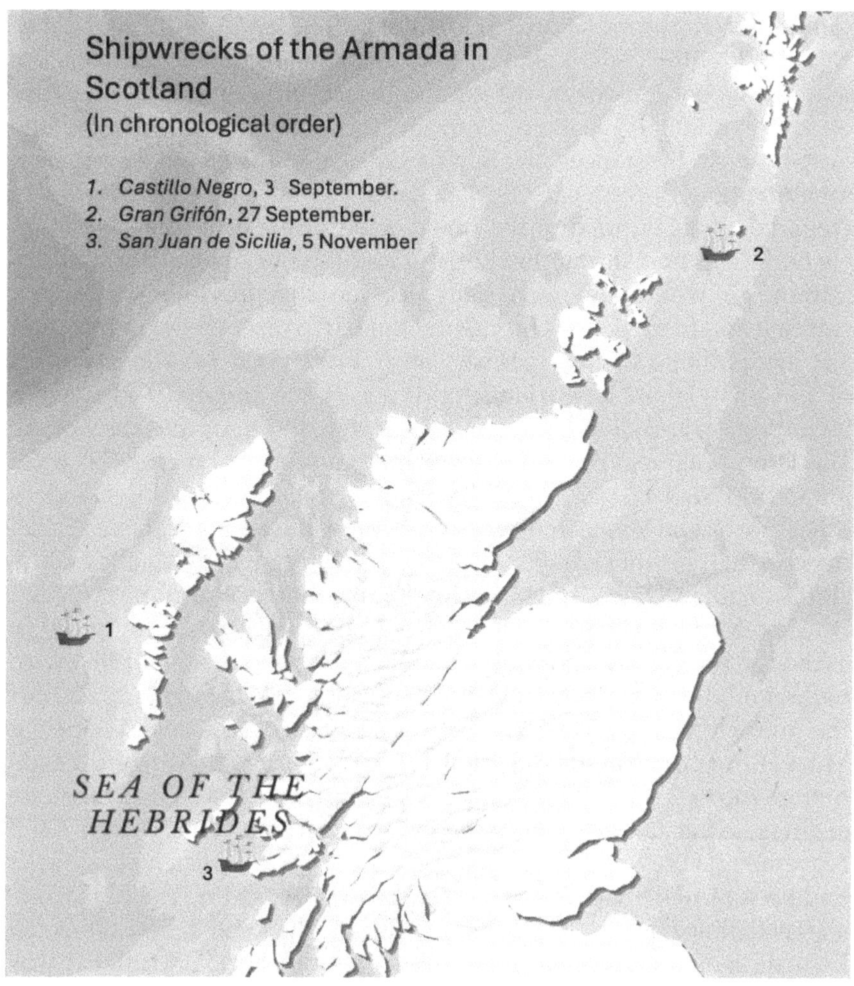

Not only Ireland served as the stage for the tragedy of the Armada. Along the Scottish coasts, three ships were also lost. At the beginning of this account, we described the sinking of the *Trinidad Valencera*, which had been sailing in the company of the hulks *Gran Grifón* and *Castillo Negro*. The former, with her holds flooded and scarcely able to navigate, concluded that the only chance of salvation was to allow herself to be carried by wind and current toward the east, in the direction of the Hebrides. Captain Gómez de Medina likely hoped to find a safe port in Norway. A similar thought may have crossed the mind of Captain Ferrat aboard the *Castillo Negro*, but that hulk was engulfed by the sea with all 200 of her crew, leaving no trace. As for the *Gran Grifón*, which had sustained significant damage during the battles of 3 and 8

August, she ran aground on Fair Isle, north of Scotland, on 27 September. Nearly the entire crew managed to disembark, except for seven sailors who drowned. However, the cold and lack of food led to the death of approximately 100 men over the following weeks. Despite the desperate circumstances, the Spaniards treated the island's seventeen inhabitants with respect until the island's owner, Malcolm Sinclair of Quendale, was notified of the unexpected presence of the castaways. He had them transported to Shetland and then to Anstruther, north of Edinburgh, where they were well received and provided with warm clothing and food. Shortly thereafter, the Earl of Bothwell – a Protestant but sympathetic to the Spanish cause – arranged for the transport of the thirty most important survivors of the ill-fated *Gran Grifón*, including several clergy, noblemen, and the captain Gómez de Medina. The latter was heavily criticised for abandoning his men, who were left to find their own way back to Spain, some travelling via Flanders, others through Denmark. Around this time, the nao *San Juan de Sicilia* docked in Tobermory Bay, within the territory of Clan MacLean. The ship's captain, Diego Téllez Enríquez, negotiated with the clan chief, Lachlan Mór MacLean of Duart, to obtain water and provisions in exchange for the assistance of 100 Spanish soldiers in his conflict with the neighbouring Clan MacDonald. Enríquez Téllez agreed, and for the following month, the Spanish forces ravaged the islands of Eigg, Muck, Rùm, and Canna. Eventually, they laid siege to Mingary Castle, though they were forced to abandon the effort due to a lack of adequate artillery. After the campaign, the soldiers returned to the ship. On 5 November, a powerful explosion rocked the bay. The *San Juan de Sicilia* had been blown apart. Two hypotheses exist regarding the cause: one suggests that Lachlan MacLean, an ally of Elizabeth I, had informed English agents in Edinburgh of the ship's presence, and that it was subsequently sabotaged by John Smollet, who planted the explosives.[91] Alternatively, according to survivor Juan de Soranguren[92] (who had already been rescued once before from the *María Juan*), the explosion was the result of an accident. Whatever the truth, the detonation killed nearly the entire crew as well as the Scottish hostages agreed upon by Enríquez and MacLean. Only sixteen men aboard survived, in addition to fifty who happened to be off the ship at the time. This group returned to Spain a year later. An exceptional case is that of the hulk *Santiago*, which, driven by the wind, anchored on 18 September near the island of Skudeneshavn (Rogaland, southern Norway), where the crew was well received. Onboard this ship – unusually for the Armada – many soldiers had brought their families, including thirty-two women and numerous children. All were eventually able to return to Spain by various means. According to Juan Gómez de Medina, captain of the *Gran Grifón*, the *Santiago* reached Norway accompanied by another hulk, whose name remains unknown.[93] This detail could help reconcile discrepancies in the number of survivors known to have sailed aboard the *Santiago*.

Perhaps the unluckiest case was that of the crew of the *zabra Nuestra Señora de Castro*, consisting of eighteen men, who, after enduring a harrowing voyage and having already sighted the Spanish coast, suddenly sank, with no possibility of saving the lives of those on board.[94] The hospital ship *San Pedro Mayor* was not much more fortunate. It had departed from La Coruña with around 150 men, but due to its role as a hospital ship, it had taken on approximately 50 wounded soldiers by the time it approached the Irish coast. After an extremely difficult journey, the ship found itself some 30 leagues from Spain when, on 6 November, a violent storm drove it into Hope Bay, England. With the vessel on the verge of sinking, the crew disembarked, only to be intercepted by an English army detachment. Although it is known that three of them were executed, the fate of the remaining crew became a matter of debate in the Royal Council, which deliberated over whether the survivors should be considered prisoners or shipwrecked sailors. The latter designation was ultimately adopted, relieving the Queen of any obligation to cover their maintenance, which was instead entrusted to the charity of the local population. The survivors of the *San Pedro Mayor* were thus distributed among the locals, and many managed to return to Spain. Those who faced the greatest difficulties were the medical personnel, who were detained due to their valuable expertise. An extreme case was that of the apothecary Lope Ruiz de la Peña, who was not granted permission to return to Spain until March 1597, making him the last prisoner of the Armada to return home.

On 21 September, the *San Martín* anchored in the port of Santander, becoming the first vessel of the Armada to return to Spain. At the end of August, she had been accompanied by ninety ships. However, successive storms – particularly those of 12, 15, and 18 September, which were especially destructive – caused several ships to fall behind or disappear entirely. Ultimately, Medina Sidonia reached the Cantabrian coast accompanied by only eleven vessels. In his letter to Philip II,[95] the Duke reported that 180 sailors had died of disease during the voyage, in addition to the sixty lost in combat, amounting to a 32 per cent mortality rate among his crew. He also requested that preparations be made to receive the crews of the ships that would continue to arrive. In his following letter,[96] he requested permission to return to his home in Andalusia to recover from an illness he had contracted during the campaign, which had brought him close to death. By the end of the month, thirty-six galleons, naos, and hulks, as well as fourteen vessels and zabras, and the *Napolitana* galleass,[97] had arrived at the Cantabrian ports of Santander and Laredo. In another letter to the King,[98] Medina Sidonia reported that several hospitals had been set up along the coast to care for 1,200 sick men, who were being fed large amounts of fruit and water. The ships themselves also required repairs, to varying degrees. None were in a condition to return to sea immediately. With Medina Sidonia incapacitated, Philip II appointed Joan de Cardona,[99] Baron of Sant Boi, as his replacement to organise relief efforts for the newly-arrived sailors and to oversee ship repairs. The King would later commission him to draft a report analysing the mistakes committed by the Armada. In various Galician ports, ships began arriving in similarly dire condition. For instance, on 7 October, the *San Juan de Recalde* docked, accompanied by the vessels *San Esteban* and *La Isabela*. Already present at that port were the galleon *San Bernardo*, the nao *San Bartolomé*, and the hulk *Sansón*, among others. There, the coordination of the operations was undertaken by Juan Pacheco Osorio, Marquis of Cerralbo, who once again demonstrated his remarkable organisational abilities. Francisco de Arriola efficiently managed the reception of ships arriving on the Guipuzcoan coast, including the *Santa Ana* of Oquendo. Despite the care and attention they received, many of the soldiers and sailors who had made it back to Spain would die in the days that followed. One of the main concerns during this time was the fate of Martínez de Leyva, prompting plans to dispatch a couple of ships to rescue him. However, this operation was never carried out, as it was literally unknown where to send them, with the reports received from survivors being contradictory. The last ship to arrive in Spain was the galleass *Zúñiga*, which would not make port until August 1589. Due to various accidents, several ships were lost on the Spanish coast, including the *Regazona* nao and the hulks *La Casa de Paz* and *Barca de Ancique*, as well as the *Santa Ana*, which was destroyed by an accidental explosion. In sum, of the 128 vessels that departed from La Coruña in July 1588, 91 returned and 37[100] were lost, representing a 28 per cent loss rate. However, beyond the quantity of ships, it is essential to consider their quality and, above all, to assess what impact this loss had on the future of the Spanish Empire.

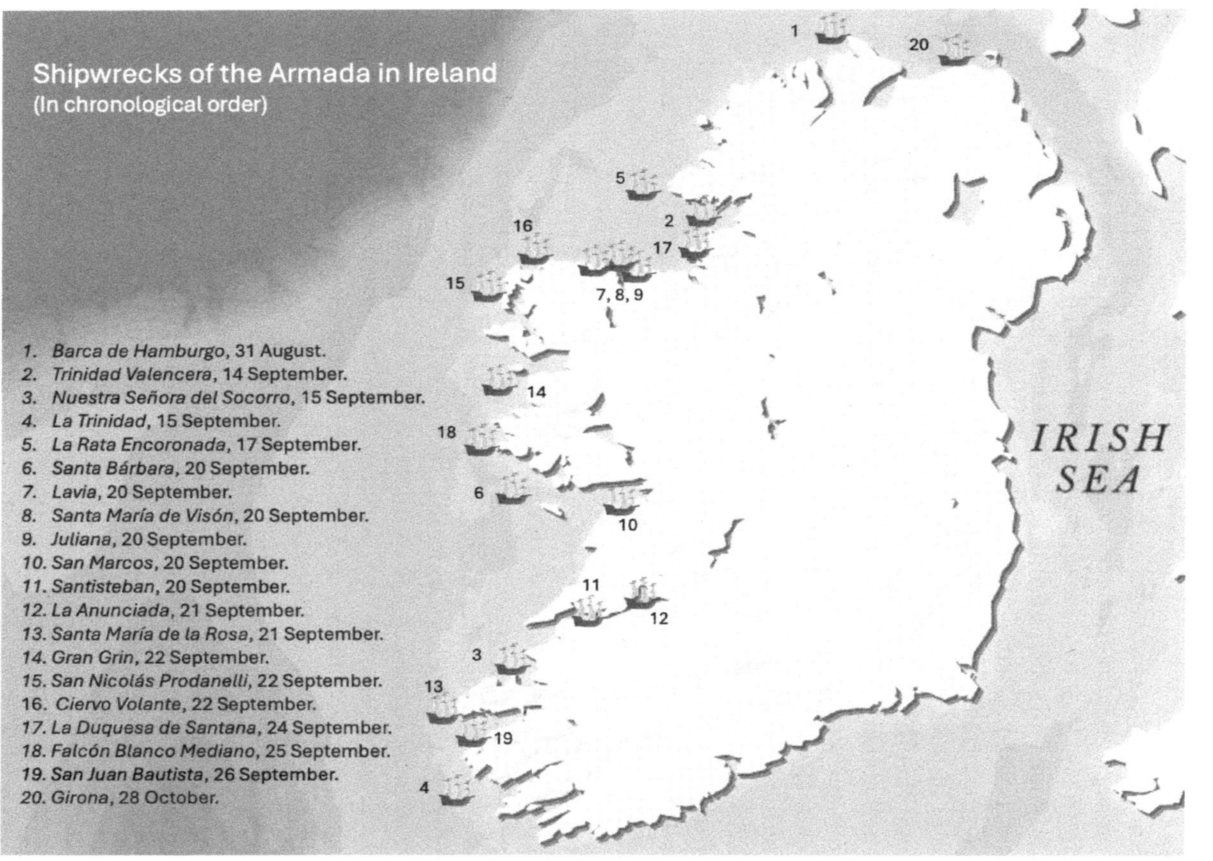

Shipwrecks of the Armada in Ireland
(In chronological order)

1. Barca de Hamburgo, 31 August.
2. Trinidad Valencera, 14 September.
3. Nuestra Señora del Socorro, 15 September.
4. La Trinidad, 15 September.
5. La Rata Encoronada, 17 September.
6. Santa Bárbara, 20 September.
7. Lavia, 20 September.
8. Santa María de Visón, 20 September.
9. Juliana, 20 September.
10. San Marcos, 20 September.
11. Santisteban, 20 September.
12. La Anunciada, 21 September.
13. Santa María de la Rosa, 21 September.
14. Gran Grin, 22 September.
15. San Nicolás Prodanelli, 22 September.
16. Ciervo Volante, 22 September.
17. La Duquesa de Santana, 24 September.
18. Falcón Blanco Mediano, 25 September.
19. San Juan Bautista, 26 September.
20. Girona, 28 October.

IRISH SEA

But the true catastrophes of war are always the losses of human life. It is practically impossible to determine an exact count of the deaths suffered by the Armada. Although it is possible to estimate approximately the number of crew members who died, we do not know who was aboard each ship at the moment of shipwreck. From the very first days of the voyage, many crew members were transferred from one ship to another, a phenomenon that became more frequent as ships encountered difficulties, already during the passage through the English Channel. However, for some ships we do possess precise figures. This is the case with the Portuguese Squadron, of which 65 per cent of the total complement of the galleons returned, despite the fact that they bore the brunt of the combat. If we examine the fate of the men aboard the *San Felipe* and the *San Mateo* – the two galleons that ran aground on the Flemish coast after Gravelines – we find that 121 men from the *San Felipe* died in combat, representing 27.56 per cent of the total crew; while the *San Mateo* lost 89 men, which constituted 22.88 per cent of its complement. These figures are not far removed from those of other galleons from this squadron that successfully returned to Spain, such as the *San Martín* (151 men lost, or 32.2 per cent), the *San Luis* (152 men, 34.62 per cent), and the *San Bernardo* (73 men, 30.93 per cent). An extreme case is that of the *San Marcos*, which wrecked off the coast of Ireland, with only one survivor out of its 386 crew members. Another figure we know is that of Armada members who were taken prisoner – in France, Flanders, England, Scotland, or Ireland. This information is thanks to the exhaustive and detailed work of Pedro Luis Chinchilla, compiled in his book *Los prisioneros de la Armada Invencible*. He estimates a total of 2,998 prisoners, of whom 1,120 were executed, 315 died of natural causes in prison (or during the harsh journeys there), and 1,558 were released following the payment of ransom. A foundational in-depth study was conducted by Calvar, Campo et al. in their monumental *Corpus Documental*,[101] in which they estimate approximately 9,150 dead and missing. Based on the general data they provide, together with studies by authors such as Gómez Beltrán and Chinchilla, I have calculated that 4,266 drowned, 777 died in combat, 1,120 were murdered on Irish beaches or executed in prisons, about 500 perished in accidents or acts of sabotage, and 653 are registered as having died from hunger and disease. This results in a total of 7,316 deaths. It is also true that among the ninety ships that returned, few, if any, arrived with their full complements. Moreover, we must also count those who died in Spain shortly after arrival, from illnesses contracted during the voyage. Thus, it is not an exaggeration to assert that the human cost of the Armada must have exceeded

10,000 men – representing approximately 35 per cent of the total. As in all wars, behind each death lies an irreparable family tragedy: many of the dead left behind widows and orphans, grieving parents, and destitute families. But in some cases, beyond these familial losses, the Empire also lost excellent sailors, soldiers, gunners – and even generals and admirals of significant value. This sombre list begins with Alonso Martínez de Leyva, Captain General of the Cavalry of Milan and one of Medina Sidonia's principal military advisors, who drowned aboard *La Girona*. Aboard *La Lavia*, Martín de Aranda, the Armada's Auditor General, drowned. Among the squadron commanders, three died in various circumstances: Hugo de Moncada, commander of the galleasses, died in combat following the wreck of the *San Lorenzo*; Juan Martínez de Recalde of the Biscayan squadron, and Miguel de Oquendo of the Guipuzcoan squadron, died shortly after disembarking from illnesses contracted during the return journey to Spain. The death of Alonso Téllez Girón, captain of the *San Marcos*, Marquis of Peñafiel and brother of the second Duke of Osuna, who drowned when his galleon sank off Ireland, was also deeply mourned at Court. It is often said that every noble house in Spain lost a relative in the Armada. While this is likely an exaggeration, many younger sons and nephews – and even some heirs – of noble families enlisted as 'adventurer knights'. Again, turning to Chinchilla's book, we find the names of several who perished. Due to his proximity to Philip II – his father being the King's Chief Equerry – Felipe de Córdoba, who died of a gunshot wound aboard the *San Marcos*,[102] is particularly notable. Also killed in action were Juan de Velasco, brother of the Count of Siruela,[103] aboard the *San Mateo*; and Francisco de Toledo, brother of the Count of Orgaz, aboard the *San Felipe*,[104] who also lost a cousin, Álvaro de Mendoza, who sailed aboard the *Trinidad Valenzera* and died en route to the prison at Drogheda.[105] This must have been a gruelling journey, during which many prisoners perished from the lack of food and water suffered in the preceding days; among the dead was Antonio Manríquez, nephew of the Duke of Nájera.[106] Aboard the *Gran Grin* travelled Ladrón Vélez de Guevara, only son of the Count of Oñate, who survived the shipwreck but was executed in an Irish prison.[107] Pedro de Andrade, son of the Count of Lemos,[108] and Francisco, son of Enrique Sarmiento, Lord of Hachas,[109] drowned aboard the *Nuestra Señora de la Rosa*. Ruy Díaz de Mendoza, the adolescent son of Antonio Hurtado de Mendoza, Admiral of the Vessels and Zabras,[110] died of illness aboard the *San Juan*. On the return journey, Alonso de Solís Manrique, son of the Lord of Ojén and Rianzuela,[111] and Antonio Manrique de Lara y Manrique de Acuña, Count of Paredes de Nava, died aboard *La Girona*. Upon arriving in Spain, Diego Pacheco – brother of the Duke of Escalona and Marquis of Villena – died, having disembarked from the *San Martín*. One of Philip II's greatest concerns was the fate of the prisoners. From the outset, negotiations were initiated with the English and Dutch authorities to secure their return, either through exchanges or ransom payments. The highest ransoms were paid for Diego de Pimentel y Toledo, Marquis of Gelves, Field Master of the Tercio of Sicily and captain of the *San Mateo*, and for Alonso de Luzón, Field Master of the Tercio of Naples and captain of the *Trinidad Valenzera* – for whom £1,650 was paid.[112] For Pedro de Valdés, Admiral of the Andalusian Squadron and captain of the *Nuestra Señora del Rosario*, £1,500 was paid.[113] As we see, Philip II and his Court showed deep concern for the fate of the Armada's survivors – a stark contrast to the neglect, bordering on cruelty, displayed by the Queen and her advisors toward their own sailors and soldiers.

Regarding the ships, how did the failure of the Armada affect the Spanish Navy? Of the 128 ships that departed from La Coruña in July 1588, 91 returned to Spanish shores – that is, 28.9 per cent were lost. This is a quantitatively significant figure, but its true impact on the future of the Spanish Empire can only be understood through a qualitative analysis. Of the thirty-seven ships lost, eight belonged to the Crown and were part of the Indies or Mediterranean Fleets, while the remainder were either loaned, rented, or requisitioned vessels. The ships belonging to the Crown were specifically:

- Galleon *San Marcos*, Oceanic Fleet.
- Galleon *San Felipe*, Oceanic Fleet.
- Galleon *San Mateo*, Oceanic Fleet.
- *Nao La Trinidad*, Indies Fleet.
- *Nao San Juan Bautista*, Indies Fleet.
- *Nao Nuestra Señora del Rosario*, Indies Fleet.
- Galleass *San Lorenzo*, Naples Galleass Squadron.
- Galleass *Girona*, Naples Galleass Squadron.

In other words, six from Atlantic squadrons and two from Mediterranean ones. Given that the theatre of operations in the war between Spain and England was the Atlantic, our analysis will focus on these fleets. In 1588, the total number of warships – between galleons and naos – was 74. Thus, the failure of the Armada resulted in the destruction or disabling of 8.1 per cent of the Atlantic fleet. Did this loss bring about the collapse of the Empire? No. In warfare, what distinguishes a decisive victory from a Pyrrhic one is the enemy's capacity to recover. I will illustrate this with two maritime examples. The first is found in one of the great myths of Spanish history: the Battle of Lepanto

(1571), in which the Spanish Navy led a coalition of Catholic states to defeat the fleet of Sultan Selim II, commanded by Admiral Ali Pasha. This was an epic victory: the Ottomans lost approximately 200 ships (70 per cent of their fleet) and suffered 48,000 casualties (dead and captured), whereas the Holy League lost only 13 galleys (4.2 per cent of its fleet) and incurred 10,000 casualties. Although this halted the Ottoman advance into Western Europe via the sea,[114] Charles of Spain was unable to consolidate power in the Eastern Mediterranean or eliminate the persistent threat of Barbary piracy. Furthermore, by the following year, Selim II's shipyards had constructed a fleet larger than the one lost at Lepanto. Thus, this major naval battle, while a resounding Spanish victory, ultimately served to maintain the *status quo* in the Mediterranean for two centuries. The second example is the Battle of Trafalgar (1805), in which Spain lost ten out of the fifteen ships of the line it had sent to engage Nelson's fleet. In this case, the victory was decisive, as Spain, already an empire in decline, lacked the economic capacity to recover.[115] By contrast, in 1589, although the seemingly endless wars faced by Philip II placed enormous strain on royal finances, the Crown still managed to recover. In fact, by that year, the Spanish monarchy had already commissioned twenty-one new vessels (sixteen galleons and five naos).[116]

Regarding the lost ships, it is noteworthy that only three were lost in combat, representing just 2.34 per cent of the entire fleet[117] – a figure that should prompt reflection among those who claim that Howard's fleet achieved a crushing victory over that of Medina Sidonia. A further 22.65 per cent were lost due to meteorological causes, while another 3.12 per cent were lost due to other factors (accidents or sabotage). Among the vessels shipwrecked on the Irish coasts, only one was a galleon – the *San Marcos*, which had fought for ten hours at Gravelines, engaging some fifteen enemy ships. The highest proportion of losses occurred among the hulk ships, of which eleven out of twenty-six failed to return. All of these were of German (six of thirteen) or Flemish (five of eleven) construction, and structurally weaker than those built in the shipyards of the Cantabrian coast.[118] The naos also suffered significantly: twelve out of thirty-eight were lost, most of them of Mediterranean origin (eight out of twelve), including the Venetian *Trinidad Valenzera* and the Genoese *Rata Encoronada*, both of which were structurally unfit for the brutal climate encountered off the Irish coast. Of the naos built in the Cantabrian region, only four of the twenty-five that set sail from La Coruña were lost. Among the *galeazas*, only one was lost during the return voyage, as these were structurally robust ships – though their rudder was a known point of weakness. Indeed, of all ships lost, approximately one-third had been constructed in Spain, while two-thirds were of foreign construction.[119] Another contributing factor to the wrecks along the Irish coast was the shortage of food and fresh water. Forced to approach a rugged, unfamiliar shoreline battered by strong winds and unknown currents, some ships ultimately ran aground or were dashed against rocks. In any case, one should not underestimate the quality, professionalism, and experience of the Spanish crews, who, in most cases, managed to return home after enduring a hellish voyage through unfamiliar and storm-tossed seas.

TABLE OF LOSSES			
Type of Ship	Returned	Lost	%
Galleons	18	1	5.56%
Naos or Carracks	38	12	31.58%
Hulks	26	11	42%
Pinnaces, Vessels, and Zabras	30	3	10%
Galleasses	3	1	25%
Total	115	28	24.35%

Source: Own elaboration.

Fleet (number of ships)	Returned	Sunk in battle	Captured	Shipwreck	Missing	Others Caueses
Portugal (12)	9		2	1		
Castille (16)	12			2	2	
Biscay (14)	11	1		2		
Andalucia (11)	9		1	1		
Guipuzcoa (14)	11		1	2		2
Levante (10)	2			6		
Urcas (21)	12			7	2	
Pataches and zabras (22)	19			3		
Galleasses (4)	2		1	1		
Galleys (4)	4					
Total (128)	91	1	5	25	4	2

Source: Own elaboration.

Chapter 5

ANATOMY OF A BATTLE: GRAVELINES 1588

The Battle of Gravelines and the retreat of the Armada toward the coasts of Scotland filled the English Court with joy, as could only be expected. There was no doubt that, at that time, Spain was the greatest global empire, its territory more than fifteen times larger than that of England. This euphoria, entirely understandable in the autumn of 1588, seems to have influenced later historians, who have drawn from those events a set of conclusions that, at least from the perspective of contemporary Spanish historiography, do not correspond to reality. Thus, it is commonly believed that the disaster of the Armada marked the turning point from which the decline of the Spanish Empire and the onset of English global hegemony both began. As seen in Davis: 'Spanish defeat marked the beginning of the decline of the Spanish Empire and made England the world's preeminent naval power, allowing the English to begin colonising North America.'[1] Beyond the claim of a turning point, it is worth noting that the Spanish Empire would continue to expand until reaching its greatest territorial extent in 1780. As for the beginning of English colonisation in North America, we will see in the chapter dedicated to the Treaty of London of 1604 that this was not the case. In the twenty-first century, we see how the internet has become the main disseminator of myths and legends surrounding the Armada. Thus, for example, on the educational website Tutor Chase, we read: 'Internationally, the defeat of the Spanish Armada marked the beginning of England's rise as a global naval power. The victory demonstrated the effectiveness of England's navy and its innovative tactics, such as the use of fire-ships.' Well, it is documented that this tactic is somewhat older; in fact, it is first recounted by Thucydides.[2] The text continues: 'It also marked a shift in the balance of power in

Europe, with England emerging as a significant player . . . The victory allowed England to continue its exploration and colonisation of the New World, which brought wealth and resources into the country.'[3] This perspective is repeated across all the websites I have consulted (more than twenty) and can be summarised in two key assertions: 'The English had superior ships, weapons and tactics' and 'It proved that England was a major naval power'.[4] Let us put this to the test. In this chapter, we will examine the four main points upon which the alleged English superiority – said to have led to the failure of the Armada's mission – is based: the ships, the cannons, the tactics, and finally, the leadership.

The Ships

There exists a general consensus in historiography affirming that English ships were smaller and more manoeuvrable than their Spanish counterparts. This purported advantage allowed commanders such as Howard, Drake, and Frobisher to handle their vessels in such a way that they could attack and withdraw at will. These ships – described by Kostman as 'of the future' – are, in his emphatic view, the reason behind the 'humiliating defeat' that, according to Robinett's triumphant account, the English inflicted. This point is crucial, as it lies at the heart of explanations for what transpired in the battles throughout the English Channel. These views have also influenced a significant portion of Spanish historiography, which has come to accept this perspective. However, this interpretation overlooks several considerations that challenge the alleged superiority of English ships. To begin with, it is important to reiterate that the Armada was not conceived as a fleet of war squadrons but rather as a combination of transport squadrons protected by a minority of warships – no more than thirty-five in total. This fact meant that, in every engagement, the English enjoyed numerical superiority. This advantage was particularly overwhelming during the crucial moment of the campaign: the Battle of Gravelines. Moreover, the English also benefitted from favourable winds, which, in a theatre of operations where sailing was the only means of propulsion, granted them a decisive advantage. Additionally, the minimal losses incurred during combat are often conflated with the ships that were wrecked along the coasts of Ireland, without considering that many of these vessels were not built in Spanish shipyards, but originated from Italian, Flemish or German ports.

As for the English ships, the reforms initiated by John Hawkins in the 1570s, in collaboration with Richard Chapman, are widely known. These reforms led to the construction of more agile and faster vessels by increasing the keel-to-beam ratio and setting the depth of the hold at two-fifths of the beam. They also reduced the forecastle and incorporated a pronounced beakhead to improve pitching, while changes to the hull structure provided greater stability. One notable result of these innovations was the *Foresight*, which Kostman refers to as 'the ship of the future' and which served as a prototype for Elizabethan shipbuilding from that point onward. Can these ships be considered the most advanced of their time? Arguably, yes – but only in the context of the commercial and military needs of late sixteenth-century England. That is, their trade was primarily limited to northern European countries, France or Spain (when not at war). These sleeker, faster, and lighter vessels came at the cost of reduced cargo capacity, which posed a serious issue on long voyages where coastal navigation was not possible. This limited their ability to carry sufficient water and provisions, a major problem during privateering missions in the Caribbean. Moreover, they suffered greatly in the harsh conditions of those turbulent waters and in regions such as the treacherous Cape Horn. As has already been noted, while privateering proved highly profitable for the English Crown (and especially for individual pirates), it is also true that it came at the cost of significant losses at sea. To give just one example: of the five ships that set out on Drake's daring circumnavigation of the globe, only one returned. Not to mention the difficulties these ships faced in open-sea engagements, which we will examine in further detail below.

What, then, of the Spanish warships? As with the English vessels, Spanish ships were designed to meet the specific needs of their empire. Spain required warships capable of navigating the trade routes across the seas, serving primarily as escorts for the treasure fleets transporting gold, silver, and goods from Spain to its colonies and back. This analysis excludes the Mediterranean fleet, which was fundamentally different from its Atlantic counterpart and consisted mainly of oar-powered vessels such as galleys and galleasses. For oceanic navigation, the ships of Philip II had to be large enough to house a substantial amount of armaments and provisions, as well as spare parts (sails, rigging, lead sheeting, ropes, anchors, etc.) necessary for carrying out repairs at sea. Moreover, since Spanish naval doctrine emphasised overcoming the enemy through elevation and boarding, ships required high forecastles and sterncastles. Consequently, it was estimated that their tonnage should be around 400 to 500 tons. Considering that during the reign of Philip II, Spain was a country deeply invested in technological innovation, can we truly believe that shipbuilders from the Cantabrian coast, the Basque Country, and Andalusia were unaware of the advancements in navigation being made by other nations? Certainly not. The improvement and development of galleons was a constant concern. From the early years of his reign, the king commissioned several studies aimed at enhancing his naval forces. In 1567, he tasked Pedro Menéndez de Avilés with the construction of twelve galleons – known as the Twelve Apostles – which would go on to protect the Spanish treasure fleet with remarkable success against the persistent threat of piracy throughout the following decade, until they were eventually replaced by more advanced vessels. These were hybrid galleons (*galeones agalerados*), propelled by both sail and oar. Their design sought to combine the advantages of traditional galleons – such as increased cargo capacity and the ability to mount artillery along the sides – with the agility and manoeuvrability afforded by oar propulsion, characteristic of galleys. This combination allowed for greater speed and better handling during critical moments in battle, without depending entirely on the wind. The ships were built with a significantly increased length-to-beam ratio (24m in length and 6.82m in beam). And while in 1570 John Hawkins launched the *Foresight* in England, in Spain the cosmographer, geographer, and seaman Cristóbal de Barros presented his *Memorial*[5] to the king. In it, he proposed a series of reforms aimed at rationalising shipbuilding, from the planting of new forests to the design of ships themselves. In 1575, Juan Escalante de Mendoza[6] published his seminal *Itinerario de navegación de los mares y tierras occidentales*, a comprehensive compendium of contemporary knowledge in astronomy, piloting, meteorology, cosmography, cartography, and naval architecture. This treatise became one of the most significant and complete works on navigation and nautical science of the sixteenth century. So valuable was its content that the Casa de Contratación in Seville prohibited its publication in order to prevent it from falling into the hands of pirates, privateers or rival nations that might exploit its wealth of information.[7] Subsequently, in 1582, by Royal Decree, Philip II commissioned Cristóbal de Barros to oversee the construction of the new galleons intended to replace those built by Menéndez de Avilés. The design and construction of these new warships generated a series of reports and counter-reports that together constitute a veritable compendium of the naval technology of the period. The new vessels were based on the operational experience of the Twelve Apostles, with Diego Flores de Valdés suggesting that their tonnage be retained, but that their beam and depth to the main deck be increased. The goal was to raise the artillery positions, thereby making their use more effective, and to reduce the height of the sterncastle.

Based on this initial proposal, two commissions were established: one in Santander, led by Cristóbal de Barros and composed of ship captains and naval constructors; and another in Seville, headed by Diego Flores de Valdés and including experienced seamen such as Diego Maldonado, Pedro Sarmiento de Gamboa, and Diego de Sotomayor.[8] Despite minor differences, both commissions reached remarkably similar conclusions. As with Hawkins' ships, both groups agreed that vessels should be lengthened and their beam reduced. In fact, the trend toward increasing the length-to-beam ratio was well known in all nations with serious interests in oceanic navigation. A Spanish aphorism vividly illustrates this principle: *Dame quilla y te daré milla*[9] ('Give me a keel and I'll give you a mile'), indicating that a longer hull favours greater speed by improving hydrodynamics and allowing for more space between masts, thus enhancing sail performance. However, in the Spanish case, such a solution had to be balanced against other strategic requirements: cargo capacity – for goods, repair supplies, and armaments; the need for prominent forecastles and sterncastles – to implement Spanish boarding tactics; and stronger construction – to withstand artillery fire, storms, and Caribbean hurricanes. This robustness, along with other measures such as lowering the gun deck, also contributed to improved stability. Thus, while English ships were designed for speed to conduct swift corsair raids and retreat with the plunder, Spanish galleons, being larger, sacrificed a small degree of speed in exchange for significantly greater resilience. Perhaps the clearest example of this trade-off is found at the beginning of the Battle of Gravelines, where, in the words of Kostman:[10]

> [Francis] Drake led the first attack in *Revenge*, followed in turn by Hawkins in *Victory* and [Martin] Frobisher in *Triumph*. At one point all three flagships lay within musket range of *San Martín*, pouring fire into her at point-blank range. During the next two hours the hull of the Spanish flagship was hit by over 200 round shots, her rigging was badly damaged and her decks were literally awash with blood. The other Spanish ships of Don Alonso Martínez de Leiva's forlorn hope were equally battered, but somehow all five galleons held the English off.

That 'somehow' was the robustness (and, I would add, the quality) of the Spanish ships – not to mention the Portuguese galleons, which had been refitted according to Spanish naval engineering standards. Indeed, if we return to the Battle of Gravelines, we find that of the twelve ships that bore the brunt of the initial assault and fought for seven hours against superior numbers, only four failed to return to Spain: *San Mateo* and *San Felipe* were severely damaged and retreated to the Flemish coast, where they were captured by the Dutch; *San Marcos* and *Trinidad Valencera* sank off the coast of Ireland. As noted earlier, the English managed to sink only one vessel during the entire campaign – the Basque *María Juan*. Furthermore, if we examine the Armada's course along the Irish coastline, we see that of the twenty-five Cantabrian naos, only four were lost; of the twenty-seven northern hulks, eleven; and of the thirteen Mediterranean carracks, eight. This underscores the high quality of Philip II's galleons, which bore the brunt of combat, often facing numerically superior forces. We must also recognise the skill and resilience of the Spanish seamen – often unjustly maligned – who demonstrated expert seamanship under adverse conditions: fighting to leeward for nearly a week in the Channel and later returning home through storm-ridden waters and treacherous coastlines largely unknown to them, navigating in the face of near-apocalyptic weather.

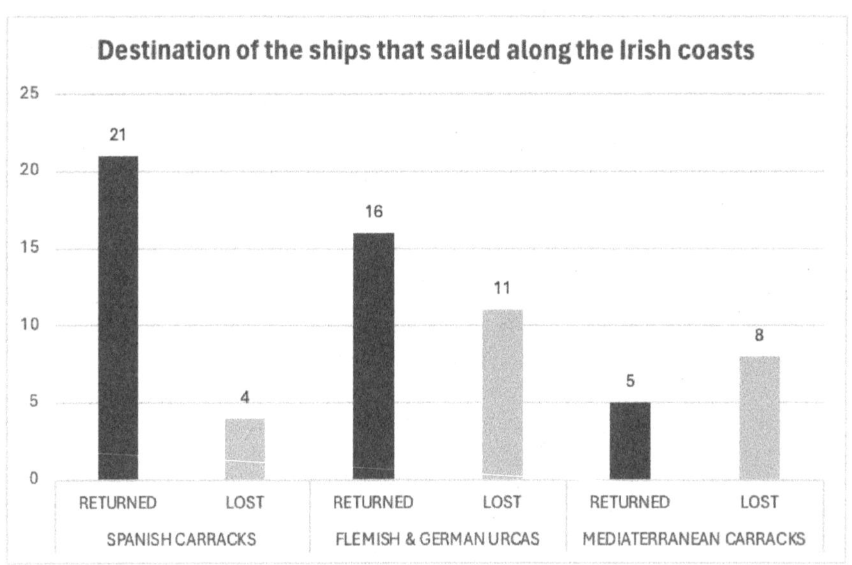

Cannons

Although it may be said that the invention of gunpowder in twelfth-century China marked the beginning of a new era in naval warfare, it would still take a long time before artillery became a decisive weapon at sea. In Europe, it was not until the fourteenth century that the first cannons were mounted on ships, yet they lacked any real tactical efficacy. Despite continuous efforts to develop more destructive artillery, naval battles in the sixteenth century were still largely decided through boarding actions. As Gómez Beltrán aptly states, maritime warfare was essentially a matter of 'boarding or avoiding being boarded'.[11] This was the approach that led to Spanish victories in the two major naval battles preceding 1588: Lepanto and San Miguel. In brief, the tactic involved three sequential phases. First, medium-range artillery was employed to immobilise the enemy vessel. In a second phase, more destructive stone-throwing cannons (*cañones pedreros*) were fired, while infantry swept the enemy deck with arquebuses and muskets. Finally, the ship was grappled using harpoons and hooks, and infantry boarded the enemy vessel. The effectiveness of this tactic relied heavily on embarking soldiers trained in the use of firearms and close combat. Thus, as early as 1537, Spain began deploying the so-called *Tercios Embarcados* (Embarked Tercios), the first to receive this distinction being the Naples Sea Old Companies, effectively the world's first marines. The value of this strategy is underscored by the fact that other European powers adopted similar forces only much later, such as in 1664, with the foundation of the English marines. However, historiography seems to have reached a certain consensus according to which 'the English seem to have surmounted the problem of working ship-killing artillery in close engagements, whereas their adversaries had not'.[12] According to this narrative, Howard's ships possessed a secret weapon – so revolutionary it could single-handedly determine the outcome of naval battles and usher in new tactics that would allow England to dominate the seas for centuries. This weapon was their new naval artillery, allegedly capable of destroying enemy ships at long range, rendering traditional boarding tactics obsolete. But was that truly the case?

Firstly, it must be noted that the importance of naval artillery did not go unnoticed by any nation, and the Spanish were no exception. To cite just one example, as early as 1575 they founded the first artillery school in the world. This was entirely logical: Spain, with possessions throughout the western Mediterranean and on both shores of the Atlantic Ocean, was naturally committed to possessing the best means to defend its territories in an unavoidable theatre of war – the sea. In any case, by the end of the sixteenth century, naval artillery was still in its early stages of development. One must begin with the fact that there was no standardisation of artillery systems. The manufacturing process for cannons was costly and production methods remained artisanal, resulting in a wide variety of artillery pieces and calibres.

For instance, Martin and Parker provide the following table of English cannons and their respective calibres:

- Royal Cannon 8in
- Cannon 8in
- Serpentine Cannon 7in
- Bastard Cannon 7in
- Demi-Cannon 6in
- Pedro Cannon 6in
- Culverin 5in
- Basilisk 5in
- Demi-Culverin 4in
- Bastard Culverin 4in
- Saker 3in
- Minion 3in
- Falcon 2⅓in
- Falconet 2in

And there were even more intermediate calibres. This variety was typical across all countries and posed a serious logistical challenge. However, it may be Angus Konstam, in his *Tudor Warships II: Elizabeth I's Navy*, who provides the clearest insight into the revolutionary nature of English cannons: 'During the 1570s, William Wynter, the Navy's Master Gunner, developed a new bronze cannon of shorter range, specifically designed for naval service, and these were in general use by the time the Spanish Armada sailed through the English Channel.' Broadly speaking, the English developed a lighter and shorter artillery piece, seeking greater mobility and requiring fewer personnel to operate. Moreover, their reduced length meant they overheated less, thereby allowing for a higher rate of fire. However, it is important to note that Spanish gunners adhered to a clear doctrine that prioritised accuracy over volume of fire. This proved especially significant during the Armada's passage through the English Channel, given that Howard's ships could resupply powder and shot at numerous ports, whereas the Spanish had only the ammunition stored in their ships' gunrooms. Nevertheless, if the situation required it – as explained in the treatise by Cristóbal Lechuga,[13] the rate of fire for large artillery pieces could range from 3.33 to 3.75 shots per hour; thus, under extreme conditions in sustained combat, it could reach a maximum of 4 shots per hour – that is, one shot every 15 minutes. For smaller artillery pieces, the same author gives a value of 6.25 shots per hour in sustained action, which is approximately one shot every ten minutes. As Gómez

Beltrán demonstrates,[14] this figure was only slightly below the rate achieved by English gunners aboard their new ships. However, it is important to note that the shorter and lighter English cannons also came with certain disadvantages. For instance, their projectiles were launched at lower velocities, resulting in less impact power due to decreased kinetic energy – unless the guns were fired at closer range. Moreover, they generated more violent recoil (which, as we will see, was further exacerbated by the type of gun carriage used), leading to greater instability and, consequently, reduced accuracy – in other words, hitting the target became considerably more difficult.[15] That is in theory – but what happened in practice with these supposedly ship-killing English cannons? Upon reading the Armada captains' reports on the battles in the English Channel, one striking detail is the sheer number of shots fired by the English fleet in comparison to the Spanish. For example, in the account of the skirmish on 31 July near Plymouth, the artillery captain aboard the *San Martín* wrote:

> In this skirmish, our flagship fired one hundred and twenty cannon shots, and the other ships in our fleet fired around six hundred projectiles; while the enemy fleet fired more than two thousand. Our flagship's fore topmast and mainstay were damaged, as well as much of the rigging; the hull sustained numerous cannon strikes. The enemy could have boarded her, but they were driven back by a volley of musket fire, and also because they dared not approach any closer than cannon range, which they managed with great speed and efficiency.[16]

In other words, the Spanish gunner was impressed by the rapid rate of English fire, but out of over 2,000 cannon shots, the reported damage was minimal. One of the clergy who accompanied the Armada, Father Juan de Victoria, stated: 'Our soldiers claim that the enemy fleet had no real fighting men, only gunners and sailors – and that the gunners were so unskilled that, out of a thousand cannonballs, hardly any hit our ships.'[17] Indeed, between 31 July and 7 August, the Armada did not lose a single ship in battle. Beyond the Spanish perception of the conflict, Martin and Parker cite a letter from William Thomas to Lord Burghley, dated 30 September, in which he remarked: 'What can be said but our sins was the cause that so much powder and shot [were] spent, and so long time in fight, and, in comparison thereof, so little harm?'[18] However, English historiography not only refers to better cannons but also to a greater number of heavy guns. According to I. A. A. Thompson, the success of Howard's fleet was not solely due to the greater number of long-range weapons, but also to the presence of short- and medium-range artillery capable of delivering heavy fire. He argues that the 330 long-range pieces (culverins and demi-culverins) aboard the Queen's galleons overwhelmed the 136 Spanish ones (culverins and demi-culverins), concluding that the English naval doctrine possessed a far more advanced conception of artillery and its tactical use than the Spanish. However, Thompson's analysis suffers from certain issues, as it overlooks important nuances. To begin with, as mentioned previously, there was no standardisation among nations regarding terminology, classifications, or, even less so, design parameters. What one navy referred to as a 'culverin' might be considered a 'medium cannon' by another; what the English classified as a 'demi-culverin' might be called a 'third' or 'quarter cannon' by the Spanish; while the English culverin roughly corresponded to the Spanish half-cannon. In both countries, the design criteria began by determining the weight of the projectile, but each used a different unit of measurement. Thus, even when the weapons were labelled similarly, they were not necessarily equivalent.[19] In this regard, once a common denominator is established by equalising for similar range and excluding swivel guns, it becomes evident that the Queen's galleons carried 645 pieces capable of firing projectiles heavier than four pounds, compared to 531 aboard the Spanish galleons – only a 17.6 per cent difference, not the 33 per cent that has often been claimed.[20] Moreover, if we focus solely on pieces with genuine destructive capability, the percentage difference is even smaller. Using the figures provided by Thompson, the following correlation of forces can be established:

	>29 pounds	27–22 pounds	21–15 pounds	Total
English Navy	3.90%	2.70%	9.20%	15.80%
Spanish Armada	2.10%	0.50%	5.60%	8.20%

In other words, heavy weaponry represented only 6 per cent of the total armament across both fleets. Moreover, if we are to believe Walter Raleigh, many of these cannons were barely used: 'So as then many of those great guns, wanting powder and shot, stood but as ciphers and scarecrows',[21] suggesting there were logistical shortcomings. Similar issues may have affected the Armada's ships, as Martin and Parker point out based on evidence recovered from the wreck of the *Gran Grifón*.[22] Thus, the proportion of cannons capable of causing structural damage to enemy vessels was 15.80 per cent in the English fleet, compared to 8.20 per cent in the Spanish. However, Thompson also fails to account for a crucial factor in artillery effectiveness: the height from which the shot is fired. As previously noted, Spanish ships were taller than their English counterparts, meaning that while Howard's galleons fired from approximately 5ft above the waterline, the Spanish guns fired from a height of around 12ft. Under such conditions, the destructive potential of both fleets' artillery was effectively equalised. Furthermore, it must be remembered that the Spanish Armada included fewer dedicated warships than the English fleet, as it was primarily a troop transport force, whereas the English fleet had been designed – at least in theory – with the specific aim of destroying the Armada. Finally, it is entirely unfounded to assert that Spanish artillery inflicted no long-range damage on the English, as Howard's reports systematically omit reference to such casualties or damages sustained during combat.

Another element Thompson neglects is the role of lighter artillery. Intermediate and light-calibre pieces (ranging from 8 to 14lbs and 7.5 to 4lbs, respectively) constituted 76 per cent of the total artillery; within this range, the advantage actually shifted in favour of the Spanish, who possessed 43.6 per cent compared to the English 32.4 per cent. These lighter pieces were also among the most frequently used, along with individual firearms which, in close-quarter engagements, could deliver a respectable impact of around 11lbs (as estimated by Thompson) per volley, to which must be added approximately 6lbs from arquebus and musket fire. That said, it should also be noted that, although some engagements were reported to have occurred at pike distance (about 20ft),[23] most were deliberately conducted beyond 55 yards in an effort to neutralise the effectiveness of Spanish small arms. These weapons, when fired in coordinated volleys, could create a deadly sweep of fire across the attacking vessel, making a sustained close-range assault virtually impossible.[24]

The second element often cited as a revolutionary technical advancement and one of the 'secret weapons' that contributed to the defeat of the Spanish was the four-wheeled gun carriage. Firstly, this was neither an invention nor a secret weapon. This type of carriage had been in use in the Royal Navy since at least 1545, as evidenced by the findings from the *Mary Rose*, which sank in that year.[25] Furthermore, it should be remembered that Philip II was King Consort of England after that date, and it is entirely reasonable to assume that either he or one of his Castilian advisors who accompanied him would have been aware of this supposed innovation. Even if that were not the case, the Spanish had captured numerous privateer vessels in the Caribbean – most notably during the 1568 expedition of Hawkins and Drake, but also during smaller operations in subsequent years. How, then, can we explain why the Spanish did not adopt this development? Simply put, it did not align with the needs of Spanish military doctrine. While the four-wheeled gun carriage had certain advantages, it also presented significant drawbacks. Among its advantages were its reduced footprint, which suited the narrower English privateer vessels, and its lighter weight and increased manoeuvrability, which facilitated quicker reloading – a feature consistent with English tactical doctrine, which emphasised a high rate of fire. However, its disadvantages were notable. Due to its reduced stability, the recoil from firing – despite the carriage being secured with ropes – caused it to jolt violently backward. This, in turn, resulted in decreased accuracy and more rapid deterioration of both the gun muzzle and the ship's deck due to impact damage.[26] In contrast, the two-wheeled gun carriage was better suited to the Spanish artillery system. Although it occupied more space, a drawback mitigated by the broader Spanish ships, its greater weight and its long trail (which acted as a lever) made it significantly more stable upon firing. Instead of bouncing, it slid backward smoothly, which enhanced accuracy – an objective explicitly prioritised by Spanish artillery regulations, which emphasised firing only when a target could be reliably hit.[27] It is also important to clarify that archaeological recoveries from Spanish shipwrecks off the Irish coast have revealed gun carriages approximately 20ft in length. Colin Martin, who has studied the wreck of the *Trinidad Valencera* extensively, explicitly[28] stated that these carriages were intended for use on land and were never part of the Spanish naval artillery system. Should we then conclude that the English were technologically more advanced simply because they employed, in the sixteenth century, the type of gun carriage that would later become globally dominant? The answer is no. This type of carriage was already known to other nations, and the reason it was not adopted was because it did not suit their military doctrine. In fact, the widespread adoption of this carriage did not occur until the second half of the seventeenth century, with the introduction of line-of-battle tactics by the Dutch, which required sustained artillery fire with the maximum possible volume.

But since paper can endure anything, let us turn to the facts: what actually happened during the Battle of Gravelines on 8 August?

Did English artillery truly demonstrate the destructive power that traditional British historiography claims? Once again, we turn to the data provided by I. A. A. Thompson, from which Gómez Beltrán[29] has calculated that if the English had 1,972 artillery pieces, and according to contemporary regulations each should have had a minimum of 20 projectiles, then – assuming 20 per cent were kept in reserve – it would mean that the English fired approximately 31,552 shots that day. This firepower was not directed at the entire Spanish Armada, but only at the roughly thirty ships that remained in the rearguard. The outcome was that by the afternoon, the *María Juan* had sunk, and the following day the galleons *San Felipe* and *San Mateo* were lost on the Dutch coast. However, these ships were lost more due to saturation fire – as the battle accounts show, the galleons in question were surrounded by over fifteen English ships for several hours – than as a result of the artillery's effectiveness, which, had it been so decisive, would have sunk them almost immediately. What is more, the *San Martín*, which fought against more ships for ten hours, suffered only thirty-two casualties (killed and wounded) and was able to return to Spain, as did the majority of the Armada vessels that fought at Gravelines. If we go back to the skirmishes of the previous days, we see that despite having more guns capable of firing at long range, the English failed to sink or disable a single ship of the Armada. In fact, of the eighteen galleons that undertook the return voyage – despite enduring violent storms and having borne the brunt of combat against Howard's fleet throughout the week only one, the *San Marcos*, was shipwrecked. This single fact alone should prompt us to reflect on the actual destructive capacity of English artillery at the end of the sixteenth century.

Tactics

Another aspect in which English historiography has proclaimed the naval superiority of England at the end of the sixteenth century lies in the field of tactics, often described as revolutionary, as for instance by Peter Brimacombe.[30] A detailed analysis of the tactics employed by both fleets will allow us to determine which side possessed a more advanced understanding of naval warfare. Let us begin with the Spanish case. One of the criticisms frequently levelled against the Spanish navy is its alleged insistence on boarding actions as the principal means of securing victory. Thus, for example, Davis claims: 'The Spanish were following the traditional but obsolescent tactics of closing and boarding, and the faster and more manoeuvrable English ships would not allow that',[31] thereby revealing a fundamental misunderstanding of Spanish military doctrine. It was not that boarding was an outdated tactic: rather, it was not the sole tactic employed. Or, more precisely, it was only a part of a far more complex military doctrine that had been formally established as early as 1530 in Alonso de Chaves' *Espejo de Navegantes*. This monumental work compiled not only Spanish naval doctrine but also all navigational routes between Spain and the Americas. The book was of such strategic importance that it was considered a state secret, and its publication was prohibited;[32] it was used to train navigators at the *Casa de Contratación* in Seville. Its fate remained unknown for centuries until a copy was discovered in a monastery and later transferred to the Real Academia de la Historia in Madrid, where it was brought to light by Cesáreo Fernández Duro in 1895. As mentioned, the *Espejo de Navegantes* set forth a three-phase Spanish naval doctrine: first, the use of long-range artillery, despite the relatively early development of such weapons in 1530; second, the approach toward the enemy and deployment of short-range artillery, along with, critically, the use of small arms – namely muskets and arquebuses. Finally, and only then, once the enemy had been sufficiently weakened by artillery and marine infantry fire, came the final phase: boarding. In other words, Spanish military doctrine was not confined to boarding, as in ancient warfare where firearms were absent, but rather fully incorporated the available military technology of the period. Let us consider two examples from the two most strategically significant naval battles of the sixteenth century. The first is Lepanto (1571) in the Mediterranean, a galley battle already discussed in a previous chapter. In this case, the decisive element was the Spanish use of arquebus and musket fire as soon as the Ottoman vessels approached,[33] whereas the Ottoman infantry was equipped only with bows and arrows. The second case is the Battle of Vila Franca (or San Miguel, 1582) in the Atlantic, involving galleons, naos and hulks, in which the Spanish defeated the French, employing not only long-range artillery but also, as previously noted, small arms and, ultimately, boarding. In both cases, the battles were decided by the strategic genius of Admiral Álvaro de Bazán, the Marquis of Santa Cruz. As Parker notes, it was a matter of boarding or avoiding being boarded.[34] In fact, during the Battle of Gravelines, the English attempted such tactics by means of a 'pack attack' (see Diagram 1), exploiting their numerical superiority. One famous episode recounts how an English sailor who boarded a Spanish vessel alone was promptly cut to pieces.

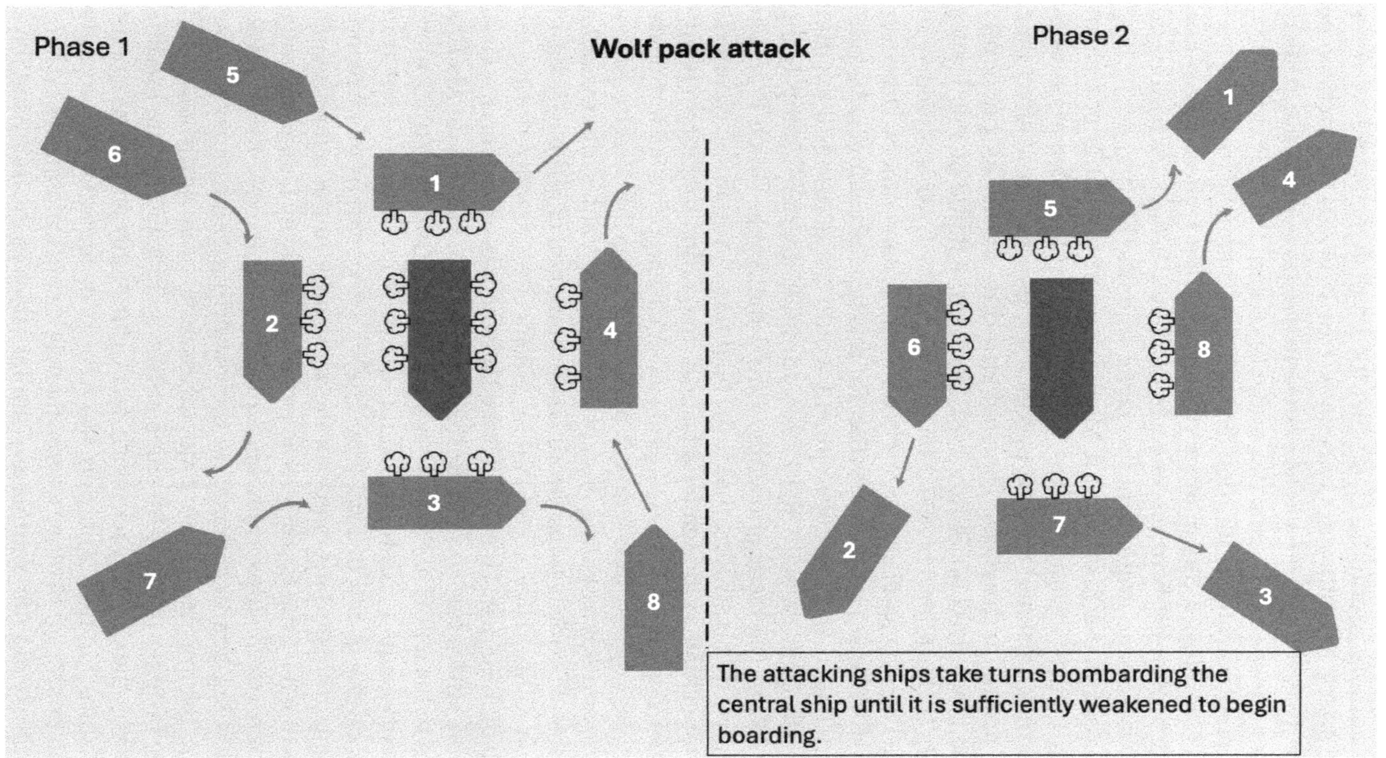

Illustration 1. Wolf pack attack.

Some authors venture to claim that the English fleet conducted the first line attack in history at Gravelines. However, for this tactic to be truly effective, it required a level of destructive power that the artillery of the time, as we have seen, did not yet possess. While it is true that at some point during the journey through the Channel, Howard's fleet may have employed this manoeuvre, it was not a tactical innovation, as it had been used before – even by an English fleet. According to Parker,[35] the first to implement this tactic were the Portuguese in 1503, when a Portuguese fleet faced a Muslim one at the Battle of Calicut, although in that instance it was a rudimentary line formation with limited artillery, aimed more at producing a psychological effect than causing actual destruction. The first true line battle, as previously mentioned, occurred on 24 July 1582, during the Battle of Vila Franca, in which the artillery of the Marquis of Santa Cruz succeeded in sinking a French vessel and severely damaging many others. However, artillery still needed to evolve further for this type of combat to become genuinely effective. That development would not occur until the mid-seventeenth century, with the Dutch admiral Maarten Tromp being the first to use the tactic efficiently during the Anglo-Dutch War. On the other hand, it is striking that some historians – usually so rigorous – resort to speculation to identify a tactical innovation that might justify the alleged English victory at Gravelines. A case in point is Martin and Parker, who suggest that Drake,[36] after capturing the *Nuestra Señora del Rosario* and interrogating its captain Pedro de Valdés, discovered a way to defeat Spanish ships through a combination of artillery fire and manoeuvrability. They argue that he put this into practice during the skirmish of 3 August. That day, Drake launched an assault on the *Gran Grifón* and struck her so heavily that her captain, Gómez de Medina, admitted in his report to Philip II that in the Channel 'they struck more than twenty shots . . . into said hulk, the principal cause of her ruin and loss on the island of Faril (Fair Isle)'.[37] However, in this instance, these eminent historians overlook the remainder of the report, which specifies that the *Gran Grifón* was not attacked solely by Drake's *Revenge*, but by ten other ships from his squadron. Moreover, the Spanish avoided being boarded thanks to their short-range cannons and the action of the infantry, who fired incessantly with arquebuses and muskets, keeping the English ships at bay. In fact, Gómez de Medina is describing the classic 'pack attack' tactic commonly employed against 'zorreros' ships[38] in the Caribbean, the Azores, or the Atlantic, typically ending in boarding. In other words, there is no tactical innovation here. Indeed, Drake did not conduct any similar attack in the following days, and, at Gravelines, his performance was, at best, questionable – as evidenced by Frobisher's remarks about him.

In the Spanish case, there was indeed a clearly defined naval doctrine, established by Alonso de Chaves' *Tratado* (Treatise), which was grounded in systematic training. As Chaves stated: 'It is said that every attempt at order at sea is futile, for no order can be maintained. To this I respond that, given equal armament, the better-organised fleet will prevail, for if the sea already confounds an orderly fleet, it will do so even more to one that does not maintain good order.'[39] This formation was based on a signalling code and a high degree of discipline in manoeuvring. In his *Tratado*, Chaves detailed how Spanish vessels were to be arranged for both defensive and offensive purposes, establishing fundamental elementary formations: in line, in squadron, and in wing (Diagram 2), from which more complex formations would be derived.

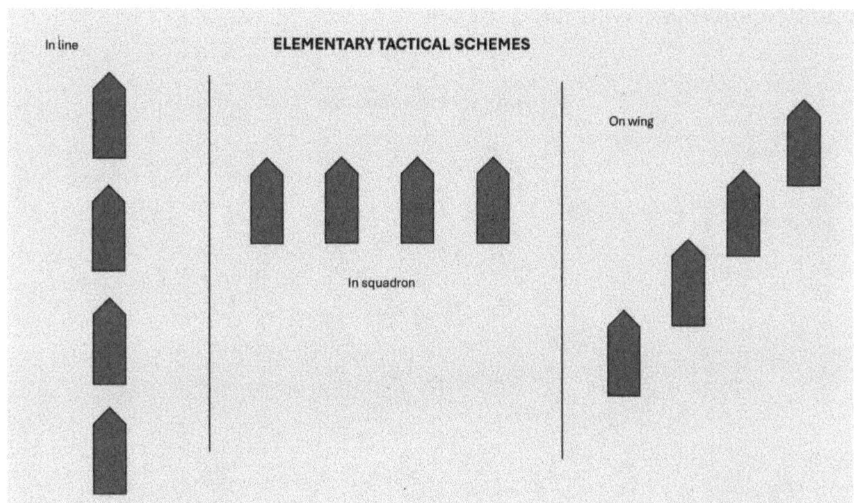

Diagram 2. Elementary tactical schemes.

In this regard, what is of particular interest for the analysis of the Armada is the *tenaza* (pincer) or 'eagle-head' formation (Diagram 3), as it is termed by Chaves. This formation was composed of three distinct components: the vanguard, the rearguard – where the warships were positioned to protect the cargo ships – and the central body, referred to as the *batalla* (or centre). The vanguard and rearguard were mobile units whose function was to defend the centre, and they could adopt various configurations in accordance with this objective. This was the system established to protect the Spanish treasure fleets bound for the Indies, and it proved overwhelmingly effective for nearly three centuries. Did the Armada employ this type of formation? Yes, because, as previously noted, it was primarily a transport fleet, in which only one-third of the ships were warships, while the remainder carried troops. Was this formation effective during the Armada's passage through the English Channel? Yes, insofar as the English were unable to break it until the attack by the fireships. Why is it commonly referred to as a 'crescent formation'? This is a misnomer. The English sailors, lacking a naval doctrine as developed as that of the Spanish, perceived the formation from a distance and believed it resembled a crescent. Thus, when Robert Adams was tasked with illustrating it, he depicted it accordingly. As for the English fleet, Adams' work makes it evident that it lacked any formalised formation doctrine – likely because, until that point, there had been no perceived need for one.

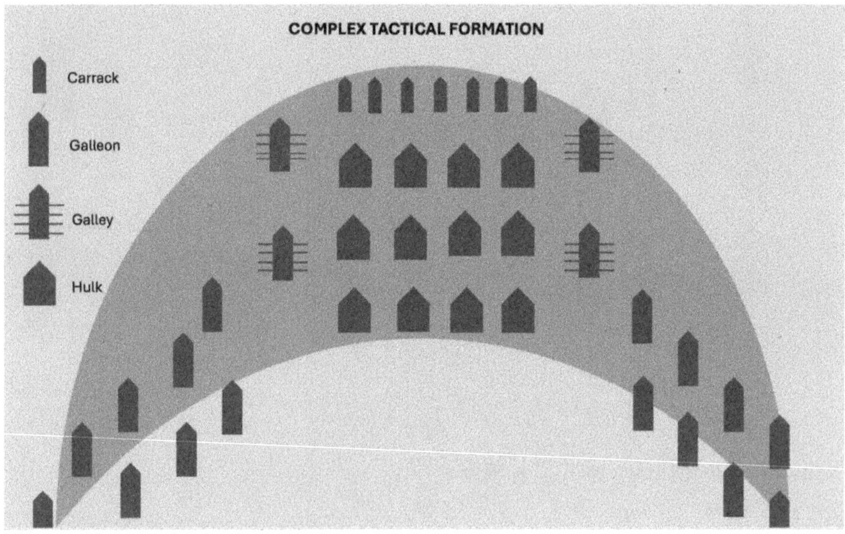

Diagram 3. Complex tactical formation

After all, the English navy of that period, aside from its commercial aspect, remained largely oriented toward privateering activities. Thus, during the early days of the campaign, they pursued the Spanish fleet in a disorderly swarm. That is, a single ship – usually the flagship – would lead the formation, which was closed at the rear by the admiral's ship, while the rest of the fleet sailed in between without any structured order, each vessel maintaining its own pace under the sole condition that none were to overtake the flagship or lag behind the admiral's ship (Diagram 4). However, on 4 August, after several engagements, it became clear that Howard of Effingham's tactical approach had been an utter failure. Despite the fact that the Spanish Armada had yielded the weather gauge, which would have allowed the English ships to manoeuvre more effectively thanks to their smaller size, Howard was unable to achieve his objective of 'plucking' the Armada by sinking or capturing the stragglers. Even when, during the action off the Isle of Wight, Frobisher attempted to flank the Armada from the north, the Spanish formation altered its course in such a coordinated manner that it maintained its cohesion throughout. As a result, at the war council held on 4 August, it was decided that the fleet would be arranged into four squadrons. This new formation, however, proved equally ineffective. As Alonso de Chaves had already explained, such a configuration was only suitable for galleys, where artillery was mounted at the prow, unlike galleons, which housed their guns along the port and starboard sides. This formation was only effective for frontal assaults and lost much of its utility in the case of galleon warfare.

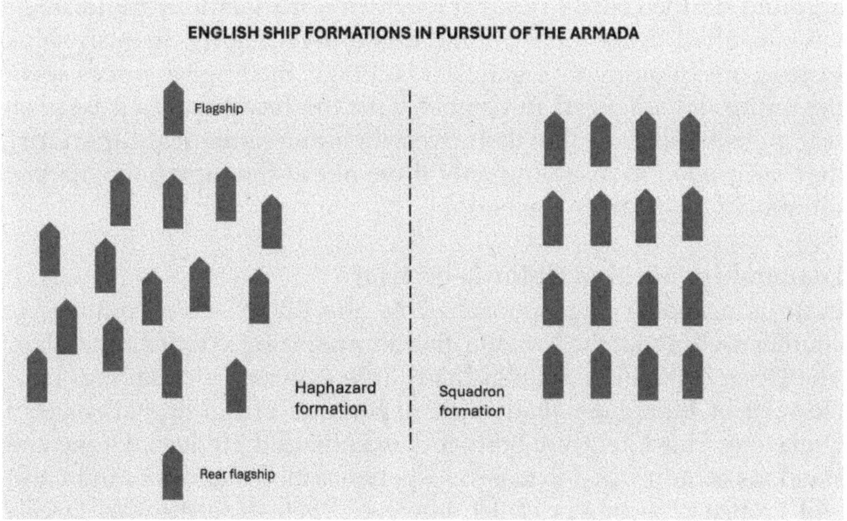

Diagram 4. English ships formations in pursuit of the Armada.

Had Howard – and with him, his admirals Drake, Hawkins, and Frobisher – possessed a sound tactical understanding, they might have pursued far more effective alternatives. For instance, by exploiting the weather gauge, they could have deployed the fleet into two wings and enveloped the Armada. However, such a manoeuvre would have required a level of coordination for which their captains, pilots and sailors were likely unprepared. In fact, when Frobisher attempted such an action, he failed to break the Spanish formation or to destroy a single ship from the Armada. Furthermore, if we analyse the behaviour of the English fleet at Gravelines, their limitations become even more evident. To begin with, Howard's inexplicable decision at the onset of the battle – to tack his entire squadron to engage the galleass *San Lorenzo* – resulted in a loss of two critical hours during which he could have overwhelmed the few galleons and warships left in the Armada's rearguard. Even during the course of the battle, the English fleet concentrated its efforts on attacking these same galleons and naos at the rear, initially with a numerical superiority of 10 to 1, and later 3 to 1. It would have been far more logical and tactically sound to allocate a portion of the English fleet to tack northward, take advantage of the weather gauge, and attack the virtually defenceless squadrons of hulks and dispatch vessels, which were scattered along the Flemish coast. These would have been forced either to confront English artillery or run aground on the coastal sandbanks. Such a manoeuvre would likely have resulted in the catastrophic loss of over eighty Spanish ships. Instead of exploiting this tactical advantage, the English forces spent the entire day engaged in combat with the heavily armed Spanish war galleons, and despite their overwhelming numerical superiority, they succeeded in disabling only three out of the forty galleons that ultimately took part in the battle.

Leadership: Medina Sidonia on trial

In an educational page published by the BBC,[40] we encounter yet another myth about the Armada, this time referring to the commanders: 'The Duke of Medina Sidonia [had] little experience of sailing. Lord Howard of Effingham [had] little experience of fighting at sea, but Drake and Hawkins were both very experienced.' Indeed, Drake and Hawkins certainly had extensive experience at sea, and we might also add Frobisher to this brief list; however, their expertise was largely confined to privateering activities. When it came to commanding a large fleet, they proved to be rather mediocre, as we shall see later on.

If we return to the BBC text, it would seem as though Medina Sidonia sailed alone, when in fact, he was supported by a considerable number of highly experienced individuals.

Perhaps the most renowned and significant Spaniard to embark with the Armada was Juan Martínez de Recalde.[41] There is a rare consensus among English historians – such as Mattingly, Collins, and Parker – who regard him as the finest Spanish seaman involved in the Armada campaign. In fact, Thompson described him as 'the nearest the Spanish could get to a sea-dog'.[42] Such unabashed arrogance! Born in Bilbao in 1540, into a family of shipowners well connected at the court of Charles I of Spain, Recalde spent his formative years learning the craft of navigation and organising fleets and courier *zabras* for the Emperor. He was soon drawn to military life and served in Flanders, initially in the infantry, then in the cavalry, and finally leading various fleets to Dunkirk in 1572, 1573, and 1575 – the latter in joint command with Pedro de Valdés. During the Desmond Rebellions in Ireland (1569–73 and 1579–83), Recalde was tasked in 1579 with transporting 800 mercenaries sent by Pope Gregory XIII to the Irish port of Smerwick, in an attempt to spark a Catholic uprising – a mission that ultimately failed. He also participated in the conquest of the Azores during the Portuguese Succession War, for which he was rewarded with a knighthood in the Order of Santiago. Around this time, he was appointed General of the Sea and married Isabel de Idiáquez, niece of Philip II's personal secretary, Juan de Idiáquez. When preparations for the Enterprise of England began in 1586, his name was one of the first to be put forward by the King and his advisors. Recalde was appointed admiral of the Squadron of Biscay, one of the Armada's principal fighting squadrons, and he participated actively in the Channel battles. During the arduous return journey, his health declined rapidly. He had contracted typhus in Lisbon in 1587 and had never fully recovered. After reaching La Coruña at the end of September, he passed away on 23 October 1588.

Miguel de Oquendo[43] represents a paradigmatic case of a self-made man in the sixteenth century. He was born in San Sebastián, likely in 1524, into a humble family. Like many young Basques, he left his hometown at the age of 13 to sail as a cabin boy to the Indies. During this formative period, he undertook numerous voyages across the Atlantic and, with entrepreneurial instincts, engaged in commerce – activities that eventually made him an extremely wealthy man. Over time, Oquendo became increasingly involved in military affairs, initially commanding only troop or supply transport vessels.

His first significant engagement was in the Battle of São Miguel (1582), under the legendary Álvaro de Bazán. In this battle – which has been discussed earlier – Oquendo commanded the rearguard, leading five Basque naos. His conduct was remarkably heroic, displaying a knowledge of naval warfare uncommon in someone fighting for the first time. He even managed to board the flagship of the French fleet, from which the admiral, the Count of Brissac, escaped by a stroke of luck at the last moment. For this feat, the King knighted him into the Order of Santiago. After participating in several smaller engagements, Oquendo's skill at sea, his bravery in combat, and his popularity and influence in Guipúzcoa made him an ideal candidate to command one of the Armada's squadrons – a position he accepted despite being over 60 years old. A proponent of a more aggressive strategy against the English fleet, he was present in all the major battles. The return voyage was extremely harsh, but his ship, the *Santa Ana*, managed to reach the port of San Sebastián. However, gravely ill, Oquendo died in early October.

If the Basque Country produced the two aforementioned admirals, the next two hailed from Asturias – a region that gave rise to a remarkable generation of outstanding seafarers such as Pedro Menéndez de Avilés, Pedro Menéndez Márquez, and Esteban de las Alas, as well as the two figures discussed below, among many others. The first of them is Diego Flores de Valdés.[44] He was born around 1530 into a family of minor nobility. By 1553, he was already serving in the Spanish Navy in Flanders, and two years later, he was sailing in the waters of Chile and Peru. In 1557, he was engaged in combat against English pirates off the Cantabrian coast. In 1565, he served as the admiral of an expedition to Florida, during which he destroyed the French colony of Fort Caroline, then serving as a base for the pirate Jean Ribault, who was captured and executed. For this, he was knighted by the King into the Order of Santiago. The following year, he was appointed Captain General of the Tierra Firme Fleet, making fourteen transatlantic voyages to the Americas between 1567 and 1584. In 1575, he was promoted to Captain General of the Royal Armada of the Indies Guard – one of the highest-ranking positions to which a Spanish naval officer could aspire. Due to his exceptional service record, both the King and the Duke of Medina Sidonia entrusted him with the command of the Castile Squadron – one of the Armada's combat squadrons – for the English campaign. A proponent of a more conservative strategy, he often managed to impose his views, consistently exerting considerable influence over Medina Sidonia. After the return of the remnants of the Armada, and while the King and the Duke sought to take responsibility for the failure, most of the officers accused Flores de Valdés of being chiefly to blame for the disaster. As a result, he was put on trial but ultimately acquitted. He spent the remainder of his life in his native Asturias, where he died in 1595.

Another of the prominent Asturian admirals was the commander of the Andalusian Squadron, Pedro de Valdés Menéndez-Lavandera – a highly controversial and polarising figure, and one of the most frequently criticised admirals among Spanish historians. He was a cousin of Diego Flores de Valdés, with whom he maintained a long-standing personal enmity. Born in 1544, Valdés volunteered at the age of 16 to serve in the galleys of the Viceroy of Naples, where an act of heroism in combat earned him knighthood in the Order of Santiago. In 1565, already reputed as an excellent military officer, he returned to Cádiz to join the Indies fleet under the command of Pedro Menéndez de Avilés, in the expedition led by Diego Flores de Valdés to eradicate the colony of Huguenot pirates established in Florida. In 1568, he married Ana Menéndez de Avilés, daughter of Pedro Menéndez de Avilés, though the marriage lasted only two months. Accused of adultery, the young bride was murdered by her father and brother, both of whom were tried and sentenced to death.[45] Thus, in the span of a few months, Pedro de Valdés lost his wife, father-in-law, and brother-in-law. He returned to Spain in 1570, carrying out various assignments for the Crown until 1575, when he was appointed Captain General of the Armada of the West and the Andalusian coasts. In 1581, he was ordered to lead the conquest of the Azores but failed due to his miscalculation of the enemy's strength. During the English campaign, Medina Sidonia entrusted him with command of the Andalusian Squadron. While in the Channel, Valdés became embroiled in a new controversy: after an accident left his ship ungovernable and adrift, he surrendered to the first English ship he encountered – Drake's *Revenge*. From that moment, he was criticised for failing to fire a single shot, for not disabling his artillery, and for not throwing overboard the 50,000 ducats on board. By preserving those assets, he likely ensured a far more comfortable captivity than his fellow officers. He witnessed the rest of the Armada's voyage through the Channel and the Battle of Gravelines from aboard the *Revenge*. After a relatively privileged captivity, he was released in 1593 and remained in Flanders by order of the King. In 1600, after the death of Philip II, his son, Philip III, recalled him to court, appointing him governor and captain general of the island of Cuba in 1602 – a post he held until his return in 1609. He spent his final years in his palace in Asturias, where he died in 1617.

We conclude with the exceptional case of Martín de Bertendona, another seasoned Basque seafarer. In the mid-sixteenth century, he had made numerous voyages through the English Channel, transporting soldiers and weapons to Flanders. From 1556 to 1571, he served in the Mediterranean, gaining extensive military experience fighting Ottoman pirates and participating in the renowned Battle of Lepanto. He later returned to the Channel, engaging in battles against the Dutch at Enkhuizen, Middelburg, and the siege of Antwerp. He also commanded a squadron at the Battle of Terceira. With such a distinguished record, he could not be left out of the English campaign. Following its failure, as we shall see in subsequent chapters, he continued to serve in the Spanish Navy with notable success.

But who was Alonso Pérez de Guzmán, Duke of Medina Sidonia – the man to whom Philip II entrusted the immense responsibility of commanding one of the largest naval expeditions ever assembled in the early modern period? Born in 1550, he inherited the title of Duke of Medina Sidonia at the age of nine – one of the highest-ranking noble titles in Spain – and possessed vast estates in Andalusia, whose revenues made him one of the wealthiest men in the kingdom. His education, as had been his father's, was overseen by Pedro de Medina, one of the most prestigious cosmographers, mathematicians and historians of his generation. His treatises on navigation, especially his *Arte de navegar* ('*The Art of Navigation*'),[46] were renowned throughout Europe. So, we must therefore assume that the young Duke acquired extensive maritime knowledge, particularly considering that the majority of his estates were located in regions whose principal economic activity was trade with the American colonies. He was an intelligent and cultured young man, and he proved to be an exceptionally effective administrator of the vast family inheritance he received – an inheritance which he not only preserved but significantly expanded. Moreover, he demonstrated unwavering loyalty to Philip II. Indeed, this loyalty led him to accept a politically advantageous marriage in 1564 to Ana Gómez de Silva, daughter of the Portuguese nobleman Ruy Gómez de Silva and Ana de Mendoza de la Cerda, Princes of Éboli. His mother-in-law was a woman of considerable influence over Philip II, albeit with a reputation – perhaps deserved – for conspiratorial inclinations. It was not uncommon for members of the House of Medina Sidonia to intermarry with the Portuguese high nobility, as the duchy held extensive estates along the border with Portugal.[47] When the Portuguese succession crisis erupted in 1580, Alonso Pérez de Guzmán played a key role in the military occupation of the neighbouring kingdom. He raised an army and successfully took control of several cities in southern Portugal.

This military success further cemented his already close relationship with the King at a delicate time for the Duke, as his mother-in-law, the scheming Princess of Éboli, had just been imprisoned, accused of orchestrating the assassination of Juan de Escobedo, one of Philip II's closest advisors. As a reward for his decisive actions, the King granted him the Order of the Golden Fleece – the highest honour in Spain – and offered him the prestigious post of Captain General of Milan. However, the Duke declined the appointment, preferring to remain close to his lands in Andalusia. Later, when the King resolved to undertake the English enterprise, Medina Sidonia quickly assumed a prominent role in the recruitment of men, ships, and provisions for the campaign. As noted earlier, he also responded effectively to the attack on Cádiz led by Drake in 1587, ensuring that the assault had no major long-term consequences for the Armada. In fact, in January 1588, he was appointed by the King as Captain General of Andalusia. Only days later, Álvaro de Bazán, the Marquis of Santa Cruz – who had been designated to command the Armada in the English campaign – died in Lisbon. Two days after Bazán's death, Philip II informed Medina Sidonia that he was to serve as his replacement. He had just been named Captain General of the Ocean Sea. What were the reasons behind the King's decision? As Thompson notes,[48] Medina Sidonia's appointment was not a surprise to his contemporaries; on the contrary, it appeared to be the most logical choice. The Duke embodied the key qualities required for such a high command: prestige, capability and wealth. Starting with prestige, it is important to understand the mindset of the period, characterised by a rigidly stratified social order. It was therefore essential that the Captain General of the Armada be a nobleman of the highest rank, one whom subordinate officers would not dare to oppose without risking charges of treason. Beyond this, Medina Sidonia was an exceptionally influential figure – not only in Andalusia but across the broader Spanish realm. He was widely admired and respected, capable of attracting nobles, sailors and volunteer soldiers to the cause. His prestige extended into Portugal as well, where the House of Medina Sidonia maintained strong family ties. His father-in-law was Portuguese and closely associated with the former royal house of Portugal. In contrast to his predecessor, the Marquis of Santa Cruz – who never enjoyed the favour of the Portuguese aristocracy[49] – Alonso Pérez de Guzmán could expect a warm reception from Portuguese nobles. Furthermore, had the Armada's mission succeeded, the Duke also possessed valuable connections in England. He had long protected the colony of English merchants established in Cádiz and had engaged in numerous commercial dealings with them.[50]

In the event that Queen Elizabeth were to be overthrown and a brief regency had to be established, there was no Spanish nobleman better connected than Medina Sidonia. Secondly, one must consider his abilities: in March 1588, with the Armada anchored in Lisbon, what was required above all was an efficient administrator capable of launching it to sea as swiftly as possible. Following the devastating epidemic that had ravaged the fleet, it was essential to impose order amidst the prevailing chaos – and at that juncture, there was no more effective administrator than Medina Sidonia. Paradoxically, as Thompson has aptly noted, the role of Captain General of the Armada did not actually entail 'leading' the fleet in a tactical sense. Admirals, ship captains, pilots, artillerymen, the *maestres de campo* of the Tercios, and even the common soldiers and sailors were all thoroughly versed in Spanish military doctrine. Each within his sphere of responsibility knew precisely what was expected of him. Moreover, it is worth emphasising that, within the royal court, Medina Sidonia was regarded as having military experience, having fought in Portugal and successfully repelled Drake's assault on Cádiz in 1587. Finally, there is the question of wealth. In this respect, Alonso Pérez de Guzmán possessed one of the greatest fortunes in Spain – an attribute of considerable importance at the time. The royal treasury under Philip II was frequently depleted, and it was often incumbent upon his generals to advance the costs of military campaigns themselves. In this instance, the Duke contributed eight million maravedís. From this perspective, there was no more suitable candidate than Medina Sidonia. Certainly, there were more seasoned naval officers or commanders – such as Leyva or Recalde – but none could rival the Duke's combination of prestige, influence, organisational capacity, and, crucially, financial means. Philip II held Alonso Pérez de Guzmán in the highest esteem and regard; were it not so, he would not have entrusted him with the hand of Ana Gómez de Silva, a young woman dearly cherished by the King, whose parents belonged to the innermost circle of his confidants. A contemporary report by Girolamo Lippomano, the Venetian ambassador, captures the sentiment held towards him in the palaces of Madrid: 'he has excellent qualities, and is generally beloved. He is not only prudent and brave, but of a nature of extreme goodness and benignity. He will be followed by many nobles, and by all Andalusia. Only one might desire in him a wider experience of the sea, but all other possible appointments presented greater difficulties.' However, Medina Sidonia sought to decline this dubious honour by every means at his disposal. In his renowned letter to the King's secretary, Juan de Idiáquez, he claimed to suffer from seasickness, to be in poor health, to lack nautical experience, and to be without the necessary funds to undertake such an enterprise. Yet the true motive was his reluctance to leave his estates in Andalusia. On this occasion, however, his complaints fell on deaf ears, as some of his excuses were rather transparently false, and, in the end, his loyalty to the King prevailed. As previously mentioned, by March 1588 the fleet was in a disastrous state, and it was Medina Sidonia's exceptional organisational capacity and efficiency that brought it to readiness for departure by the end of May – a logistical feat that has not always received due recognition. Once embarked, and following the necessary pause in La Coruña, the Armada set sail for England. Up until that point, his life had been far more immersed in administration than in naval command. Nevertheless, the decisions Medina Sidonia made during the voyage through the Channel were not taken in isolation, but carefully weighed with the counsel of his advisers. The pertinent question is: were these decisions the correct ones? Given the outcome, the simplest answer might be no. Yet a more nuanced analysis may yield a different perspective. The first major decision he faced was whether to attack the English fleet anchored at Plymouth. At first glance, this seemed an unparalleled opportunity to destroy a significant portion of the enemy's ships – or so argued Recalde and Oquendo. In the end, however, Medina Sidonia heeded the advice of Diego Flores de Valdés, who urged him to adhere to the original plan and seek to unite with the Duke of Parma's forces. Subsequent historians have criticised Medina Sidonia for failing to capitalise on this opportunity. It is indeed probable that the Spanish galleons could have inflicted serious damage on the English fleet, especially given the limited manoeuvrability of Howard's ships, which would have facilitated boarding actions. Yet, it is equally true that the Spanish may have lost some of their own warships, which were essential for the success of the mission. Furthermore, although around 100 vessels were docked at Plymouth, they did not represent the entirety of the English fleet, which was scattered across various southern ports and reinforced by Seymour's squadron at the far end of the Channel. To this must be added the tactical imprudence of expending gunpowder and ammunition at the very outset of the Channel crossing. Whereas the English had access to sixteen supply points, the Spanish had no secure harbour until reaching Flanders. Ultimately, the most compelling reason for abstaining from attack lay in the King's explicit instructions: not to destroy the English fleet, but to cross the Channel with the Armada intact in order to join with the Duke of Parma's army. Undoubtedly, it cannot have been an easy decision. Moreover, refraining from the attack meant ceding the windward position to Howard. Nevertheless, judging by the subsequent events, one might argue it was the right decision, since Howard's fleet failed to prevent the Armada – aside from a couple of damaged ships – from traversing the Channel with relative ease. The second decision for which Medina Sidonia has faced criticism was his choice to await the Duke of Parma's fleet at Calais. This highlights one of the fundamental weaknesses in the King's plan: the lack of an operational naval base along the Flemish coast. On 7 August 1588, what options were truly available to Medina Sidonia for rendezvousing with Parma's forces? The two Flemish ports under Spanish control – Dunkirk and Nieuwpoort – were surrounded by sandbanks, rendering them inaccessible to the Armada's deep-draught vessels, or at the very least posing serious risk of grounding. A similar challenge would have arisen had they attempted to reach Antwerp, due to the necessity of ascending the Scheldt River. The best options – Ostend and Flushing – were held by the Dutch. Another possibility was to head directly for Dover and disembark the infantry. However, it would have been highly unrealistic to expect to overcome the English army with merely 6,000 troops, even if they were drawn from the formidable Spanish Tercios. The Armada would have found itself caught between a rock and a hard place, and the Spanish infantry stranded in hostile territory, with no guarantee – as indeed proved to be the case – that an uprising among English Catholics would come to their aid. In short, such a move would have risked a still greater disaster. Ultimately, Calais presented itself as the best of a poor set of options: exposed to the sea and prevailing winds, yes, but with sufficient depth to accommodate the entire fleet. Though not an allied port, it was at least neutral, the primary concern being its strong and unpredictable currents.[51] Ultimately, Medina Sidonia had been given unequivocal instructions: to make contact with Parma and, should that prove impossible, to return with as many ships as possible. Calais, without doubt, represented the least unfavourable of the available options. The third major decision he faced was the selection of the route for the return journey. Once again, Recalde and Oquendo advocated for the more aggressive alternative: to retrace their path through the Channel and engage once more with the English fleet. Such a course of action would have been bold, yet it entailed two significant challenges. First, the fleet would have had to sail close-hauled for the entire journey, a manoeuvre that hindered the navigation of large ships, considerably slowing their progress (especially considering that, for example, the hulks were already notably sluggish vessels). Moreover, the galleons and warships had by then depleted a substantial portion of their ammunition. This raised the serious risk that they might find themselves unable to defend against renewed attacks mid-voyage, rendering them defenceless and vulnerable to the English fleet, which had the advantage of numerous ports for resupply. The persistence of these admirals in proposing this course of action suggests that, even after the Battle of Gravelines, the Spanish fleet was not so enfeebled as to preclude further engagement. Indeed, as we know, the Armada repeatedly challenged Howard's fleet during the subsequent pursuit along the Scottish coast – challenges to which the English commanders did not dare respond. Neither was seeking refuge in Scandinavian ports a feasible alternative: these were not allies, and the presence of such a considerable fleet would have posed both financial and diplomatic difficulties for Philip II. Thus, no viable course remained but to return by circumnavigating the British Isles – a seemingly more prudent option, with a higher likelihood of preserving a greater number of vessels for the return to Spain. Despite the violent storms and shipwrecks along the Irish coast, this course may rightly be regarded as the least damaging available. Beyond these strategic decisions, it would be unjust to reproach Medina Sidonia for any lack of military resolve or courage during the engagements in the Channel. The *San Martín* was consistently positioned at the forefront, rushing to the aid of stricken vessels and engaging the English fleet – frequently outnumbered – displaying the bravery and mettle expected of the Captain General of the Armada. Upon the fleet's return, a scapegoat was required. Naturally, it was unthinkable to criticise the original plan, as this would have been tantamount to blaming the King himself.

Accordingly, there was no shortage of voices who sought to lay blame upon Medina Sidonia – most notably the Duke of Parma.[52] Others at court similarly criticised him for his 'lack of experience and resolution',[53] citing his failure to aid Pedro de Valdés, his decision to abandon the anchorage at Calais, and the fact that he was among the first to return to Spain, allegedly leaving the greater part of the Armada to its fate. Nevertheless, it was the Duke of Parma who came under the most severe criticism.[54] It is also true that the prevailing sentiment was one of absolving Medina Sidonia of all responsibility, as is clearly demonstrated by two of the principal reports drawn up to assess the failures of the Armada.[55] In summary, the conduct of the Duke of Medina Sidonia was, in general terms, beyond reproach. While some of his decisions may be open to debate, it remains the case that he adhered faithfully to the King's plan, succeeded in organising the Armada at Lisbon in record time, and managed to lead the fleet through the entirety of the English Channel without losing a single ship in combat – at least until the Battle of Gravelines. As for the return voyage, he chose the most prudent course available, and the subsequent shipwrecks caused by adverse weather cannot fairly be attributed to his actions.

In conclusion, it must be stated that neither English ships nor their artillery were revolutionary. Rather, they were simply adapted to meet the commercial and military needs of the English fleet. Their innovations were not unprecedented discoveries made by the Queen's seamen. When comparing vessels of equivalent size – such as the *Ark Royal* and the *San Martín* – we find no demonstrable superiority in speed or manoeuvrability. This widespread misunderstanding is likely fuelled by a desire to impose an epic narrative upon the failure of the Armada. Yet it overlooks the fact that the Battle of Gravelines was but a single episode in a protracted conflict that would endure for 16 years. Events following that battle call into question the triumphalist conclusions to which some historians have hastily arrived. During the passage through the Channel, were English ships faster and more agile than their Spanish counterparts? Yes. But was this due to their being more advanced? No. In general terms, English vessels were smaller, and thus more easily handled; they carried lighter loads – unlike the Spanish, whose ships had to be stocked with all that was required for both combat and a full-scale landing in England. The English, by contrast, could rely on multiple ports for resupply and thus sailed comparatively unburdened. Moreover, the Spanish fleet adhered strictly to a disciplined military formation which, though limiting manoeuvrability, had proven effective in the past and would remain so in future engagements. The English, under Howard, possessed no such constraints and enjoyed greater tactical freedom. With regard to artillery, the available data does not support the claim that either fleet possessed an intrinsically superior system. Rather, each had developed weaponry suited to its respective strategic and operational requirements. English vessels, designed to be lighter and faster – with a narrower beam – were necessarily fitted with smaller artillery pieces that could be accommodated within their limited internal space. Furthermore, English naval tactics favoured a high rate of fire over accuracy, resulting in a strategy that prioritised volume of shot, even at the expense of precision and efficient ammunition use. By contrast, Spanish ships, built to endure long and hazardous Atlantic crossings to the Americas, ranked cargo capacity and structural stability. Their broader design allowed for larger artillery pieces, which, in turn, facilitated a preference for fewer but more accurate and powerful shots. Thus, by the late sixteenth century, both nations had developed artillery systems that produced roughly equivalent results, without any clear technical superiority on either side. Indeed, without artillery capable of decisively determining the outcome of naval engagements, battles in 1588 – and well into the seventeenth century – continued to be resolved through boarding actions. From the accounts of Spanish captains, we know that the English admirals did attempt to board, and that they failed only because Spanish warships proved impervious to Howard's fleet. As for formation, while Medina Sidonia's ships sailed in the famously impenetrable 'eagle' configuration, the English fleet shifted from an uncoordinated structure to one resembling a phalanx – a formation better suited to galley fleets in the Mediterranean than to engagements in the Channel, as the events at Gravelines would demonstrate. Regarding leadership, we have already established that Medina Sidonia was the appropriate man for the role. He was not a seasoned seaman, but he was the capable organiser the Armada required at a moment of extreme disarray, following the devastating plague that had swept through Lisbon. Naturally, he made mistakes – just as Howard and Drake did, as previously discussed. But the issue did not lie in the Armada's leadership. The Duke was continually advised by highly experienced naval and military officers. In short, the Spanish failed in their objective of landing in England and overthrowing the Queen. Yet the root causes of this failure are not to be found in any inherent English naval superiority. Rather, they stem from a plan that was virtually impossible to execute, given the insurmountable challenge of linking the Armada with the army of the Duke of Parma. If we examine the outcome of the Battle of Gravelines in purely objective terms, it appears far closer to a stalemate than to a resounding victory for Elizabeth I's fleet.

Chapter 6

BEYOND 1588

One of the key challenges we face in understanding the true significance of the defeat of the Armada in 1588 is the lack of proper contextualisation. In other words, the failure of the Enterprise of England must be seen as part of a broader conflict that continued until 1604. It is truly disappointing to observe how some prominent British historians have treated the phase of the war spanning from 1589 to 1604, approaching it from a superficial and highly partial perspective. For instance, Parker summarises the subsequent war by stating that 'the English captured or sank Spanish ships and goods every year and damaged Philip's possessions, culminating in the capture and sacking of Cádiz in 1596', or when he presents the English Armada of 1589 as if it had been a success: 'In May 1589, Sir Francis Drake returned to Galicia at the head of a large Anglo-Dutch fleet. He put ashore an expeditionary force – something the Spanish Armada had failed to do – that first sacked Corunna and then anchored near the mouth of the Tagus while English soldiers marched on Lisbon. Although the troops soon turned back, their audacity infuriated the king [Philip II]',[1] an incomplete perspective that is also repeated in the works of other distinguished British Hispanists, such as Henry Kamen.[2] All this merely serves to reinforce the myths and half-truths that continue to surround the 1588 Armada. In the following pages, I will outline the most significant events that took place between 1589 and 1604, with particular emphasis on the various naval expeditions launched by both countries, culminating in the critical campaign of 1597.

But first, we must begin with the English Armada of 1589. While Medina Sidonia's ships were steering toward the Scottish coast, Elizabeth saw an opportunity to strike at the Spanish Treasure Fleet near the Azores. However, the condition of the ships and their crews rendered such a mission impossible. The English vessels were in urgent need of repairs – a tacit admission that the Battle of Gravelines had come at a substantial cost. As Gorrochategui indicates,[3] England was, objectively, as mired in tragedy as Spain. As a result, they had to wait until the following year to launch a far more ambitious expedition: to decisively defeat Philip II. To that end, a series of bold objectives were established, listed in a letter from Lord Burghley dated 20 September.[4] Had these been achieved, they would have displaced Spain as the dominant world power, replacing it with England:

* Destruction of the remnants of the Armada: Scattered across various ports along the Cantabrian coast, particularly in Santander, and potentially extending as far as Seville. This would not only eliminate the remaining Spanish war galleons, but also sever the vital link between the Iberian Peninsula and the Americas.
* Secession of Portugal: By conquering Lisbon, the aim was to provoke a nationwide uprising against Philip II and place António of Crato on the throne. In doing so, Portugal would become a satellite state under English influence.
* Conquest of the Azores and capture of the Treasure Fleet.

To this end, Elizabeth selected the two men she considered to be her finest commanders: Francis Drake was to lead a fleet of 180 ships, organised into five squadrons under the command of the following admirals (along with their respective flagships): Francis Drake aboard the *Revenge*; John Norris on the *Nonpareil*; Thomas Fenner, vice-admiral, aboard the *Dreadnought*; Roger Williams, second-in-command of the army, on the *Swiftsure*; and Edward Norris[5] on the *Foresight*. John Norris would also command an army of 16,000 men (115 companies, organised for the first time in English history into regiments). Altogether, the expedition comprised 27,667 men, including both sailors and soldiers, according to a document by Lord Burghley cited by Gorrochategui.[6] However, a provisional figure of 23,375 men is commonly accepted.[7] This made it the largest Armada assembled in the Atlantic Ocean during the entire sixteenth century.[8] The enormous costs were borne by António of Crato (who participated in the expedition alongside his son Manuel and numerous Portuguese supporters), Queen Elizabeth I, and, predominantly, by private investors – as was customary in such enterprises. Adverse weather conditions delayed the departure of the English Armada by several weeks. This resulted in the consumption of much of the stored provisions aboard the ships, which would prove to be one of the contributing factors to the ensuing disaster. The fleet finally set sail from Plymouth on 13 April 1589, and from the outset, the objectives laid out by the Crown began to be overshadowed by the interests of private investors. This shift in priorities is the only way to understand why Drake's fleet, rather than heading to Santander – where most of the surviving ships from the Spanish Armada were undergoing repairs – sailed to La Coruña, where only five ships were anchored. Drake believed a great treasure was being kept there, which ultimately dictated his decision. On 4 May, approximately 5,000 men disembarked under the command of John Norris. They easily laid waste to the suburb of Pescadería,[9] but upon reaching the city walls, they encountered fierce resistance from Spanish soldiers and, notably, from the citizens of La Coruña themselves – among them María Pita, a woman who would later become a legendary heroine. After 15 days of combat, Norris's forces were forced to withdraw, having suffered disproportionate losses, including two ships and 400 of their finest soldiers.[10] This failed assault also gave the Spanish and Portuguese two critical weeks to strengthen coastal defences in anticipation of further attacks. On 26 May, the English Armada reached the coast near Peniche, roughly 60km (around 40 miles) north of Lisbon. There, a war council was held in which Drake and Norris engaged in a heated debate. Drake advocated for a direct naval assault on Lisbon, while Norris favoured landing at Peniche and marching inland, believing – based on assurances from António of Crato – that the Portuguese population was eager to rise up against the Spanish and support their cause. Drake rightly argued that such a plan was folly, as the troops lacked wagons and mules and would be forced to march long distances on foot while carrying provisions and ammunition.[11] Ultimately, Norris prevailed and landed around 12,000 men at Peniche. Their initial objective was to capture the local castle, defended by approximately 1,000 troops. It took two days to subdue the garrison, and the campaign began under grim circumstances: the soldiers, already demoralised by their failure at La Coruña and the costly battle at Peniche, now faced a 40-mile march with dwindling supplies and the daunting task of taking the Portuguese capital. Throughout the march, the English forces were met with apathy from the local population and constant harassment by Spanish-Portuguese light cavalry. Meanwhile, Drake's fleet anchored at Cascais rather than advancing into the Tagus estuary (the Mar de Palha), where an eighteen-galley fleet under Martín de Padilla and Alonso de Bazán lay in wait.[12] This decision is striking, especially considering the widely asserted naval superiority of English galleons over Spanish galleys – and the overwhelming numerical advantage, roughly 150 ships[13] to 18. However, as this work has consistently shown, Drake, ever the privateer, rarely engaged in battle unless he possessed the element of surprise. Eventually, Norris reached the outskirts of Lisbon. The days that followed were a veritable nightmare for the Queen's infantry: by day, they endured relentless bombardment from Lisbon's artillery and Padilla's galleys; by night, the feared commando-style raids of the Spanish Tercios, known as *encamisadas*, inflicted mounting casualties and further eroded morale. Meanwhile, Drake had abandoned Norris and his men to their fate and remained anchored at Cascais, engaging in what he did best – seizing merchant vessels. Most of these were slow, heavy hulks from neutral northern European countries, transporting foodstuffs to Lisbon for trade. These vessels were plundered for provisions, only to be returned and compensated after the expedition's failure. Finally, after several of his regiments were virtually annihilated, Norris ordered a retreat during the night of 4 to 5 June. The remaining forces marched toward Cascais, but fewer than 4,000 of the original contingent that had departed from Peniche[14] reached the coast – many had perished, been captured, or deserted. By mid-June, the looming threat of a new Spanish fleet from Andalusia (which would raise the total number of Spanish galleys to twenty-eight) convinced both Drake and Norris that the time had come to return to England. The English infantry reembarked on 18 June. By then, food scarcity had become a critical issue, and the diet of the vast majority of soldiers and sailors was reduced to consuming raw wheat captured from Hanseatic hulks.[15] Under such conditions, starvation and scurvy soon ravaged the English Armada's crews. In addition, the lack of wind left the English ships virtually immobilised, which allowed the Spanish galleys to launch attacks, sinking or capturing between nine and eleven ships, not including smaller vessels. A few days later, off the Galician coast, an unexpected storm sank an undetermined number of ships and dispersed the rest. Drake managed to regroup part of the fleet, but the situation was dire. As Hume writes,[16] 'the men died, and were thrown overboard by the hundred from scurvy, starvation, and wounds'. On some ships, scenes of utter horror unfolded: aboard the *Gregory*, only eight healthy men remained; on the *Bark Bonner*, merely two men and two adolescents were still fit for duty.[17] Drake then ordered a landing at Vigo, at the time a small, undefended village, which was attacked by the 2,000 troops still available, who devastated everything in their path. Afterward, the fleet was divided: Norris would return to England with one portion, while Drake, taking command of the best ships and remaining sailors, set course for the Azores in a final attempt to intercept the Spanish Treasure Fleet. However, within a few days, unfavourable winds pushed Drake's fleet back toward England. They arrived in Plymouth on 10 July – only to find that Norris's ships had not yet arrived. They would not reach port for another three days. If, in the case of the Spanish Armada of 1588, it was the men who found themselves without ships – broken by the storms they endured – then in the case of the English Armada, it was the ships that returned without men. Those vessels that did make it back were manned by barely a handful of able-bodied crew. Hume estimates[18] that no more than 5,000 men returned, although this figure does not take Spanish documentation into account. According to English payroll records from September, 102 ships returned to England, and only 3,722 men claimed wages.[19] If we include knights and personal servants (who were unpaid), this would align with Hume's estimate. In short, the number of dead, captured, and missing from the English Armada was nearly double the losses suffered by the Spanish Armada the previous year. Moreover, the sickly sailors returning to Plymouth triggered a devastating epidemic that decimated much of the local population. To compound matters, Drake and Norris each blamed the other for the expedition's catastrophic failure. Despite their poor relationship, Drake and Norris agreed to conceal the full extent of their failure from the Queen, initially misleading her about the outcome of the expedition. Shortly thereafter, Norris wrote a carefully worded letter to Sir Francis Walsingham, informing him of the events that had transpired. However, he insisted above all that 'for the honour of Her Majesty and the reputation of our country, I think no fault should be found in our actions.'[20] The general's message was met with astonishment in London, but propaganda is yet another weapon of war – one in which the English were true masters. Thus, it was decided not only to avoid acknowledging the failure of the expedition, but to portray it instead as yet another resounding victory over Spain. The Elizabethan propaganda machinery was set into full motion. As a result, we have works such as *A True Coppie of a Discourse Written by a Gentleman Employed in the Late Voyage of Spaine and Portingale*, attributed to Anthony Wingfield, which offered the triumphalist narrative the political context demanded. Nevertheless, there were consequences for those involved. First and foremost, the failure marked the end of António de Crato's aspirations to the Portuguese throne. He would spend the remainder of his life drifting between the courts of London and Paris, pleading for a renewed military expedition. Eventually, he settled in Paris in early 1594, where Henry IV granted him a modest pension, barely sufficient to sustain him with dignity. Crato continued to dream of launching another campaign to reclaim Portugal, but this never came to pass. He died on 26 August 1595, sick and impoverished, in a humble residence in the Marais district. As for Norris, he emerged relatively unscathed from the debacle and would soon be entrusted once again with military command in the ongoing conflict with Spain. The same cannot be said of Drake, who fell into disgrace. He would not lead another expedition for five years – and that final venture would also be his last.

With the failures of both Armadas, the year 1590 did not mark the end of the war; in fact, it had only just begun. Both nations emerged weakened, yet Spain had the wealth of the Americas to recover economically. Thanks to these resources, Philip II embarked on an enormous naval construction programme. He had understood that he could no longer rely on embargoes or the leasing of ships from private shipbuilders; the Crown itself would have to oversee the construction of galleons for its exclusive use. By the summer of 1590, Spain already had 100 new ships sailing the seas, amounting to a total of 48,200 tons.[21] What other nation could have accomplished such a feat? This programme was accompanied by the fortification of key strongholds both in Spain and across its Caribbean territories. In London, it was well understood that the Spanish Treasure Fleet (*Flota de Indias*) was the cornerstone of Spanish power, and from that point on, Elizabeth's primary obsession became its interception. With this goal in mind, she launched major expeditions each year to locate, attack, and seize as many treasure-laden ships as possible. All of them failed. The only consolation came in the capture of so-called *zorreras* – ships that sailed alone and unprotected by the main fleet, often Portuguese in origin. One of the most persistent figures in these efforts was George Clifford, 3rd Earl of Cumberland, who personally organised a total of twelve expeditions. The most successful of these took place in 1592. Near the Azores, Clifford's fleet coincided with another expedition under the command of Martin Frobisher.[22] Though they did not encounter the *Flota de Indias*, they did capture the unprotected Portuguese carrack *Madre de Deus*, a vessel of 1,600 tons. The English ships surrounded her like a pack of wolves, bombarding her until the captain surrendered. The cargo was extraordinary: spices, silks, carpets, cloth of gold, pearls, ivory – an overwhelming collection of luxury goods valued at approximately £500,000 (equivalent to roughly £85,823,650 today). However, the successful expedition ended on a bittersweet note. Its arrival in London caused a sensation. The citizens of the capital could scarcely believe their eyes. Everyone wanted to visit the *Madre de Deus*, witness its unimaginable riches – and take a souvenir home. Ironically, the 'pirate queen' had her own loot plundered in her home port by her own subjects. The division of the remaining spoils also led to significant conflict. Initially, Cumberland and Sir John Burgh, commander of the infantry in the Raleigh-Frobisher expedition, were told they would receive nothing. Their vehement complaints eventually secured them modest compensation – though Burgh considered it an insult. The dispute ultimately led to a duel with John Gilbert (probably a relative of Sir Walter Raleigh),[23] which resulted in the death of Burgh. In the end, the Crown was able to collect the not insignificant sum of £100,000[24] (equivalent to £17,164,730 today), an amount that provided some relief but did not resolve its serious financial troubles. In other ventures, Clifford gained far more modest profits, such as in 1589 and 1593.[25] However, in most cases, his expeditions resulted in significant economic losses. For instance, the 1594 campaign[26] could have made him one of the wealthiest men in England when he captured the Portuguese carrack *Cinco Chagas*, a vessel of no less than 2,000 tons – the largest ship of its time. Yet during the attack, the ship caught fire and sank. A subsequent letter from the Venetian ambassador revealed that the *Cinco Chagas* had been carrying the most valuable cargo ever shipped from the East Indies, estimated at 2,000,000 ducats (£134,453,800 in today's currency). Had the fire been avoided, it would have constituted the greatest treasure ever captured. His fortunes were even worse in 1591, when his expedition of five ships was ambushed by six galleys under Admiral Francisco Coloma in what is now known as the Battle of the Berlengas Islands (off the Portuguese coast, approximately 60 miles north of Lisbon). The English lost three ships and suffered an undetermined number of casualties.[27] Speaking of naval engagements, we must also note that in the same year, the Battle of Flores took place in the Azores. In this engagement, a fleet commanded by Thomas Howard, Earl of Suffolk, was tasked with intercepting the Spanish Treasure Fleet near the Azores. The Queen herself became involved in the ambitious enterprise, lending six of her galleons, notably the *Defiance* (the flagship, captained by Howard), the *Revenge* (serving as rear-admiral under Vice-Admiral Richard Grenville), the *Nonpareil*, and the *Bonaventure*, along with eighteen additional ships. Upon learning of this plan, Alonso de Bazán assembled a formidable fleet of fifty-five vessels and 7,200 men, including sixteen royal galleons. He launched the pursuit of Suffolk's fleet with five squadrons, commanded respectively by Pardo Osorio, Marcos de Aramburu, Antonio de Urquiola, Martín de Bertendona, and Luis Coutiño. When the English sighted the Spanish formation, they turned and fled toward English ports. Yet one of the Queen's galleons lagged behind – or perhaps stayed behind to challenge the Spanish fleet, awaiting his compatriots to return and support him, or even sacrificed itself to buy them time. This is a question we may never be able to answer definitively. That ship was none other than the *Revenge*, under the command of Richard Grenville, who, in a seemingly suicidal or reckless act, chose to stand and fight. Over the course of several hours, the Queen's galleon – formerly Drake's flagship in numerous campaigns – resisted four successive attacks by the Spanish fleet. At last, during the night, with nearly the entire crew dead or wounded, including Grenville himself, the legendary English vessel surrendered. It was surrounded by some survivors of the 1588 Armada, including Admiral Martín de Bertendona, who commanded the galleon *San Bernabé*. Alonso de Bazán, a man of honour, ordered Grenville to be transferred to his own ship, the *San Pablo*, where efforts were made to treat a gunshot wound to the head, but his life could not be saved. Thus ended the life of one of the greatest English mariners of the century. The *Revenge* was to be towed to Spain, but in one final act of defiance, she sank in the middle of the Atlantic – choosing, perhaps, to rest on the ocean floor rather than serve her enemy.

Several weeks later, the *Flota de Indias* arrived safely in Spain, unloading its vast riches into the eager coffers of the Spanish Crown. Equally unsuccessful were England's attempts to circumnavigate the globe once more, such as the ill-fated 1589 expedition led by John Chudleigh[28] (also known as Chidley), whose ships were swallowed by the treacherous waters of the Strait of Magellan. Another failure came in 1591 with Thomas Cavendish's second expedition, which aimed to establish a trade route between England and the markets of China and Japan. Commanding a fleet of six ships and 350 men,[29] Cavendish attacked and razed the city of Santos in Brazil, where he remained for several weeks, thus missing the optimal seasonal window to cross the Strait of Magellan. When he eventually attempted the passage in late January 1592, the expedition was struck by disastrous weather: some ships were lost, others deserted and returned to England. Cavendish himself chose to pursue a route to China via the Cape of Good Hope, but he died under mysterious circumstances during the voyage. In 1593, Richard Hawkins – son of the renowned John Hawkins – set out with two ships under the pretence of conducting a geographic exploration, though his actual aim was to raid the Pacific coast. This time, the Strait of Magellan was successfully navigated, but his fleet was detected by a Spanish squadron. After a fierce battle, Hawkins surrendered under the condition that he and his men be repatriated to

England. Admiral Beltrán de Castro accepted the terms, and Hawkins later reported in his travel account: 'after our surrender, the general took especial care for the good treatment of us, and especially of those who were hurt'.[30] However, the Spanish Inquisition intervened and demanded their imprisonment. After some negotiation, the inquisitors succeeded in transferring the captives to a prison in Seville. Despite Castro's continued efforts to secure their release, Richard Hawkins did not return to England until 1602.[31] In 1595, Walter Raleigh departed from Plymouth with five ships and five pinnaces in search of the mythical city of El Dorado. He returned empty-handed but penned a sensational and exaggerated account of his expedition that captivated European audiences. The historian Martin Hume would later describe the work as a 'coarse fabrication', written primarily to entice investors in London's financial sector to fund a new voyage. Despite its impact, Raleigh's narrative failed to secure even a single shilling for his cause. These repeated incursions finally exhausted the patience of Philip II, who authorised Carlos de Amézola (also known as Amézquita) to execute a long-devised plan to attack English coastal settlements. On 26 July 1595, four galleys under the command of Diego Brochero departed from the base at Blavet and landed approximately 400 Spanish soldiers in Mount's Bay, Cornwall. Spanish troops had set foot on English soil. Within hours, the towns of Mousehole, Newlyn, Saint Paul, and Church Town were sacked. In Penzance, the local fort was assaulted, prompting the flight of a garrison of 1,200 soldiers and militiamen. On 4 August, Spanish forces transferred the fort's cannons to their galleys and began fortifying their position in anticipation of a retaliatory force of 8,000 troops en route from Plymouth. However, upon learning that an English fleet had already sailed from Plymouth, they avoided entrapment by reembarking swiftly and successfully evaded interception by English vessels.

What Amézola could not have anticipated was that the English ships sent in pursuit were, in fact, part of a much larger expedition being organised by Francis Drake and John Hawkins for an assault on the Caribbean.[32] The two privateers, having resolved their long-standing disputes, had joined forces once more. Drake, who had regained Queen Elizabeth's confidence, hoped to execute a historic campaign: to finally intercept the Spanish Treasure Fleet, devastate the Spanish Caribbean, and establish a permanent English base in Panama. For this ambitious endeavour, the expedition was equipped with twenty-eight vessels, including six of the Queen's own galleons – *Defiance*, *Garland*, *Hope*, *Bonaventure*, *Foresight*, and *Adventure*.[33] These ships carried a formidable force of 1,500 sailors and 3,000 soldiers, the latter placed under the command of Thomas Baskerville. It was the largest fleet England had ever dispatched to the Caribbean. The expedition began poorly: its primary objective – to intercept the Spanish treasure fleet – was thwarted from the outset, as inclement weather had delayed the fleet's departure from San Juan, Puerto Rico. Short on provisions, the English commanders opted for an initial assault on the Canary Islands, anchoring off Las Palmas on 4 October. There, they landed 500 soldiers using 47 barges. Thomas Baskerville had boldly promised to take the city within four hours. However, the city's governor, Alonso de Alvarado y Ulloa, was fully prepared for the attack. As the English troops reached the shore, they were met with a relentless barrage of cannon, musket and arquebus fire, forcing the survivors to retreat back to their ships within minutes. Baskerville then revised his promise, claiming he would take the city within four days. Yet Hawkins and Drake, unwilling to prolong the failed assault, set sail for the Caribbean, arriving at San Juan on 22 November. Coincidentally, the *Nuestra Señora de Begoña*, a galleon under repair and commanded by Sancho Pardo Osorio – then head of the treasure fleet's military squadron – was docked in the city. At that time, the city was guarding thousands of silver ingots valued at 3,000,000 pesos (equivalent to £64,089,130 in current prices). Upon learning of this, the English prepared to assault the city. However, on both attempts, they were repelled with heavy casualties. The defensive strategy ordered by Philip II and brilliantly executed by the engineer Bautista Antonelli had proven effective. English losses were estimated at around 400 men, while the Spanish lost only 40. It was during this campaign that John Hawkins died at the age of 63. He had been a privateer, slave trader, smuggler, and trusted agent of the Queen, who appointed him Treasurer of the Navy and tasked him with the renowned reform of English galleon design. During the 1588 Armada campaign, he ranked third in the English command hierarchy. The exact cause of his death remains uncertain – some attribute it to illness, others to wounds sustained during the San Juan assault. Drake, never one to favour attacking well-defended positions, withdrew from Puerto Rico on the night of 25 November and set course for Cartagena de Indias. But upon seeing Antonelli's newly designed fortifications, he passed the city by. He then resolved to pursue the original goal of the expedition: the conquest of Panama and the establishment of a permanent base. On 6 January, the fleet anchored off Nombre de Dios. The town's governor, Alonso de Sotomayor, aware of the port's indefensibility, retreated inland with both civilians and military forces to defend Panama City. Baskerville led his troops into the interior but quickly fell into a deadly ambush, losing over 500 men. Upon hearing of the disaster, Drake ordered a return to Nombre de Dios, which was then burned in retaliation. At this point, morale within the expedition had collapsed. As a last resort, the English hoped to encounter a galleon or carrack to seize. However, they now faced an even more pressing concern: the dire lack of food and fresh water. Drake, deeply demoralised and suffering from dysentery, died on 28 January 1596, at the age of 55. His body was consigned to the waters of Portobelo Bay. With Drake's death, England lost its most famous privateer, and Spain, its most feared pirate. Drake had been a bold and astute seaman, yet he was often entrusted with operations that far exceeded his military capabilities. As Kamen aptly described, 'Drake was by mentality a pirate rather than a general'.[34] The expedition had become orphaned – it had lost both of its admirals, fifteen captains, and twenty-two officers. The remaining leaders resolved to return to England. However, near the Isle of Pines (modern-day Isla de la Juventud, Cuba), the English fleet encountered a Spanish squadron under the command of Bernardino de Avellaneda, consisting of eight galleons and thirteen auxiliary vessels dispatched specifically to pursue them. In a state of panic, the English forces prepared for flight. In their desperation, they jettisoned all non-essential equipment, including their cannons, and only thus were they able to elude their pursuers. Ultimately, only eight of the twenty-eight ships that had set sail the previous year managed to return to England. By coincidence, at the very moment the survivors disembarked at Plymouth, the Spanish treasure fleet was arriving in Sanlúcar de Barrameda with one of the largest cargos in its history – 20,000,000 pesos (equivalent to £427,229,846 in current prices). Just days prior, *Nuestra Señora de Begoña*, under the command of Sancho Pardo Osorio, had already arrived with the silver from Puerto Rico.

In 1596, rumours began to circulate in England that a new Spanish Armada was being prepared to invade the country – and they were not unfounded. For years, a group of Irish exiles led by the Archbishop of Tuam, James O'Hely (Gaelicised Seamus Ó hÉilidhe), alongside nobles John de Lacey and Maurice FitzGerald,[35] had been pressuring Philip II to launch an Armada in Ireland, capitalising on the fact that part of the island had risen in armed rebellion against the English. Eventually, they succeeded in persuading the monarch, and by the spring of 1596, a fleet was already being assembled in various ports across Spain and Portugal.[36] In response to this intelligence, Elizabeth opted for a pre-emptive strike, ordering Charles Howard of Effingham to organise a naval expedition akin to Drake's 1587 raid. Thus, on 13 June 1596, a new English Armada set sail from Plymouth. It consisted of between 100 and 120 Anglo-Dutch ships[37] – though Spanish sources report as many as 150[38] – divided into four squadrons under the command of Effingham himself, Thomas Howard, Walter Raleigh, and Francis Vere. In total, the fleet carried 15,000 men, including 1,000 veteran Dutch soldiers, under the leadership of Robert Devereux, Earl of Essex.[39] The English were also aided by secret intelligence on Spanish defences provided by Antonio Pérez, a former secretary to Philip II who had fled Spain after orchestrating the assassination of a senior court official. A week later, the fleet sailed past Lisbon and, seeing that the city was well defended by Diego Brochero's fleet, decided to continue toward Cádiz, where, according to Pérez's reports, the defences remained as weak as they had been in 1587. In the early hours of 30 June, they arrived to find between forty-three and fifty ships and eighteen galleys anchored in the bay, tightly packed in a defensive formation that initially thwarted English assault plans. However, the Spaniards faced a critical weakness: none of their admirals were present in the city, save for Luis Alonso Flores, commander of the Tierra Firme fleet. His ships, owned by private merchants, were nearly ready to sail to the Americas with a cargo valued at 4,000,000 ducats. Under pressure from the shipowners, Flores agreed to depart immediately, inadvertently breaking the defensive line and allowing the English to enter Cádiz in a ferocious assault. The fighting lasted five to six hours and was brutal. The Captain General of Andalusia, the Duke of Medina Sidonia, was only able to muster around 5,000 poorly armed and inexperienced militiamen, who arrived too late to prevent the city's Major, Antonio Girón y Zúñiga, from surrendering. For 15 days, the English plundered Cádiz at will, while Medina Sidonia failed to organise an army or fleet capable of expelling Effingham's forces. After two weeks, the English departed, having taken hostages and looted warehouses, palaces, private homes, and churches. Before leaving, they set fire to 290 houses – about 20 per cent of the city – and sailed away. The total damage to Cádiz was estimated at 5,000,000 ducats, while the loot seized amounted to £122,231,660 in current value. Nonetheless, as both English and Spanish historians concur, 'in strategic terms, the expedition proved to be a costly failure'.[40] For instance, the English made no further attacks on other ports, even though another fleet lay anchored in nearby Sanlúcar, and they did not attempt to intercept the Treasure Fleet. While the victory was spectacular and a number of London merchants made astronomical profits, the Queen was left perplexed: her orders had not been followed. The bulk of the Spanish war fleet remained anchored in Lisbon and the ports of the Cantabrian coast, and the Treasure Fleet remained untouched.

But if a Queen was left stunned, a King was left furious. Philip II was beside himself with rage. He immediately summoned Martín de Padilla y Enríquez, Count of Santa Gadea, appointing him commander of the Armada stationed in Lisbon, and urged him to prepare the fleet for an offensive against Ireland.[41] His rear-admiral would be Diego Brochero, while the land forces would be commanded by Sancho Martínez de Leyva – whose brother had drowned off the Irish coast in 1588. A formidable Armada was outfitted, comprising 126 ships divided into six squadrons, led by, among others, Martín de Bertendona and Pedro de Zubiaur.[42] Once again, everything was ready for departure in the early days of autumn – a season notoriously unsuitable for navigation in Irish waters. Yet, in a surprising change of course, Philip II decided at the last moment that the Armada would not target Ireland but Brittany, with the initial objective of capturing the port of Brest.[43] The fleet departed from La Coruña on 25 October, but an unforeseen and massive storm struck off Cape Finisterre, devastating the expedition. The ships were forced to return to the Galician ports – but not all made it back. Between twenty-five and thirty-two ships were reported sunk or missing, including three war galleons: the *Santiago*, the *San Jerónimo*, and the *San Felipe y Santiago*. A total of 1,706 men were lost. On this occasion, Philip II did not lose his composure. He ordered the damaged ships to be repaired and new galleons and naos to be constructed in preparation for another expedition in 1597. However, the Spanish king faced a significant problem: it was not merely that his financial resources were finite, but so too was his capacity for borrowing. The situation had reached a critical point, and on 29 November, Philip II was compelled to declare a suspension of payments[44] with the aim of forcing a renegotiation of the Crown's debt and securing major debt reductions. In his favour was the knowledge that many of the Italian bankers involved – seeking to raise additional capital to lend to the Spanish Crown – had established a secondary market for Spanish debt involving Dutch and German financiers. These arrangements placed the bankers themselves at risk of financial collapse, in a situation eerily reminiscent of the 2007–08 American subprime mortgage crisis. Their own fortunes were now at risk of being dragged into bankruptcy. Thus, negotiations proceeded, and once again Philip II was able to avert disaster. Nevertheless, the suspension of payments did not come without consequence. It impacted the wages of soldiers stationed in Flanders, sparking mutinies and the loss of several strategic positions – strongholds that had cost the Duke of Parma great bloodshed to secure only a few years prior.

1597 marked the final year in which large-scale naval expeditions were launched. Both Spain and England were preparing new armadas. Elizabeth's fleet set sail with the objective of destroying the Spanish Armada stationed in the Galician ports and attacking the treasure-laden *Flota de Indias*. On this occasion, 120 ships were prepared under the command of Sir Robert Devereux, Earl of Essex, as Admiral and General-in-Chief; Sir Thomas Howard, Earl of Suffolk, as Vice-Admiral; and Sir Walter Raleigh as Rear-Admiral, in what became known as the 'Islands Voyage'. An additional twenty-five Dutch vessels joined the expedition under the command of Jacob van Wassenaer Duivenvoorde. However, the capricious god Aeolus turned his fury upon the Anglo-Dutch fleet. Shortly after their departure from the port of Plymouth, a violent storm erupted, sinking an undetermined number of ships and scattering the rest. A small portion of the fleet returned to England, while another group managed to regroup and continue their expedition toward the Galician coast. By September, they were sighted off the coast of Lisbon, triggering a state of high alert across all Spanish and Portuguese ports. During this time, a pinnace brought word to Raleigh that the fleet observed anchored at Ferrol had departed for the Azores, presumably to serve as an escort for the Spanish Indies Fleet. The English admirals convened immediately and set course for the Azores. By mid-October,[45] a new Spanish Armada – comprising 136 ships and over 10,000 men – set sail for Falmouth. It was now or never. This moment represented the most critical juncture of the war – truly a turning point. Simultaneously, the largest Spanish Indies Fleet ever assembled – comprising forty-three ships and carrying 10,000,000 pesos[46] (equivalent to £213,630,433 in current prices) – was en route to the Azores from the Americas. And Essex was heading to meet it. Meanwhile, a Spanish Armada was approaching a defenceless England. In the event of an English victory, the spoils would not only bring millions of pounds sterling to the Crown but could also provoke the definitive collapse of Spanish imperial power. Conversely, in the event of a Spanish victory, Elizabeth's reign could be imminently jeopardised – if not extinguished altogether. The tension must have been at its highest. At last, Essex's fleet arrived in the Azores, where it attacked the city of Vila Franca on the island of São Miguel. Meanwhile, the immense *Flota de Indias* had anchored in the city of Angra, on the island of Terceira. Its commander was Admiral Juan Gutiérrez de Garibay, a seasoned mariner with an impeccable record, who had already triumphed over English ships at the Battle of Pinos in 1596. Upon receiving intelligence of Essex's presence, Gutiérrez de Garibay convened his officers: should they engage an enemy fleet of more than 120 ships, or attempt an escape, taking advantage of the favourable winds? The Spanish commanders, exercising prudent judgment, chose the latter. The fleet weighed anchor and sailed at full speed towards Spain. Essex, distracted by operations on São Miguel, had been outmanoeuvred. And Philip II had been spared. However, there remained the Armada under Martín de Padilla. As it approached within approximately 27 leagues of the English coast, a sudden storm swept the fleet southward, scattering it in all directions. On this occasion, the ocean was more merciful to the Spanish: only two galleons and one hulk were lost; the rest of the ships survived. Quoting the chronicler Cabrera de Córdoba, Fernández Duro reported that seven vessels reached Cornwall and disembarked 400 soldiers, who fortified their position in anticipation of reinforcements. However, only a pinnace arrived – bringing the order to return to their ships and sail back to Spain. Both major operations of that year ultimately ended in failure. Weeks after his return to England, Essex presented himself before the Queen to voice complaints regarding Raleigh's lack of discipline. A grave misstep on the part of Robert Devereux – perhaps unaware of the deep affection Elizabeth harboured for his dashing rival. That moment marked the beginning of the end of Essex's political and military career.

From that point onward, large-scale military operations came to an end. In Spain, concern mounted over the alarming state of the king's health: Philip II was dying. He had never enjoyed robust health, but his physical condition had deteriorated significantly – especially after receiving news of the death of his daughter, Catherine Michaela. At the age of 30, she died during the birth of her eighth child.[47] She had been one of his most beloved daughters, along with Isabel Clara Eugenia, both born from his marriage to Elizabeth of Valois – the woman he had loved most profoundly in his life. There was now an urgent need to bring at least one of the ongoing wars to a close. The most promising front for reaching an agreement appeared to be with France, under the cunning and pragmatic leadership of Henry IV. To understand the context of this proxy war over the preceding decade, it is necessary to briefly consider the internal situation in France. Since 1589, the French political landscape had been in constant and unpredictable flux. That year's events had already taken a dramatic turn in December, when the former French king, Henry III, ordered the assassinations of the Duke of Guise and his brother Louis, Cardinal of Guise. The French Catholic faction never forgave him. In August 1589, a Dominican friar assassinated Henry III. The throne then passed to

Henry of Navarre – who took the name Henry IV – despite being a Protestant. Civil war persisted in France, now with the Catholic League backing Charles of Bourbon, son of Louis de Condé, as their claimant. This led to the First Expedition of the Duke of Parma to France, launched with the goal of lifting the siege of Paris laid by the forces of Henry IV. In August, in a *blitzkrieg avant la lettre*, Parma entered Paris in triumph, prompting the retreat of the Huguenot army. Meanwhile, a Spanish fleet of 37 ships and 6,470 men[48] under Admiral Sancho Pardo Osorio launched an assault on the coastal settlement of Blavet (modern-day Saint-Louis), securing a strategic beachhead for the continued conquest of Brittany, then in Protestant hands. The port also provided Spain with a potential base of operations for attacks on England.[49] Alarmed by these developments, Henry IV of France urgently appealed to Queen Elizabeth for aid. Aware of the strategic threat posed by a Spanish naval base on England's doorstep, she promptly organised a new expeditionary force. Soon after, Charles of Bourbon died, leaving the Catholic League without a clear candidate to champion. In response, Philip II, by mid-1591, resolved to send Farnese once again to France – this time with the aim of imposing his daughter, Isabel Clara Eugenia[50] (a granddaughter of Henry II of France), as monarch. Parma accepted the assignment reluctantly. The situation in the Low Countries was deteriorating week by week. The Spanish army was overstretched, fighting on too many fronts, while the Anglo-Dutch coalition began to achieve strategic victories at Zutphen, Deventer, and Steenwijk. Nevertheless, in April, the Duke of Parma successfully lifted the siege of Rouen (Normandy), facing a combined Anglo-French force led by Henry IV and Robert Devereux, Earl of Essex. He subsequently captured the town of Caudebec, though, as was often the case, he exposed himself to great personal danger and was wounded in the arm. Parma later secured further victories, defeating Anglo-French forces at Craon and again at Ambrières, where the enemy was under the command of Sir John Norris. As a result, both Brittany and Normandy began to tilt back toward the Catholic cause. In May, Parma made a second triumphant entry into Paris. He returned to Flanders soon after, but his health had been irreparably compromised. The wound sustained at Caudebec never fully healed and eventually became gangrenous. On 2 December 1592, Parma died at the age of 47, leaving Europe in mourning. He was widely regarded as one of the finest generals of the sixteenth century – perhaps the very best. His death marked the loss of a truly irreplaceable leader. As for Henry IV, faced with a war he could not win and a Catholic League unable to agree on a suitable pretender, he made a shrewd political move: in July 1593, to the astonishment of many and to the outrage of Elizabeth I – who felt deeply betrayed – he converted to Catholicism. Ever the pragmatist, Henry IV famously remarked: *'Paris vaut bien une messe'* ('Paris is well worth a Mass'). However, the conflict with the Huguenots continued. A significant episode occurred in 1594 during the battle for Fort Lion,[51] located on the Crozon Peninsula. The fort was defended by 400 Spanish troops under the command of Captain Tomé de Paredes. Strategically vital, the fort commanded a large portion of the Breton coastline. In October, an Anglo-French army of 6,000 men under the Huguenot Baron de Molac and Sir John Norris laid siege to it, while a fleet led by Sir Martin Frobisher attacked from the sea. Weeks passed, and the defiant tercios continued to hold out despite the decimation of the garrison – including the death of Captain Paredes. With his own forces suffering heavy casualties, Norris called upon Frobisher to lead a desperate assault through one of the breaches. During the attack, Frobisher was severely wounded. The following day, Norris approached the shattered bastion under a flag of truce to negotiate terms. The only surviving Spanish officer received the delegation, but once inside, the English troops treacherously turned on their hosts, killing the last remaining defenders. Only thirteen soldiers of the Tercios managed to survive, fleeing the massacre and subsequently receiving protection from the French, who were scandalised by the dishonourable conduct of John Norris. Anglo-French casualties are estimated at around 3,000 men[52] – although Norris significantly downplayed this figure in his own highly embellished and self-aggrandising account, which prioritises fantasy over factual accuracy. Among the dead was the French Marshal Yves du Liscouët, while John Norris himself was wounded. However, the most consequential loss was that of Sir Martin Frobisher, who succumbed to his injuries a few days later. England thus lost one of its most capable seamen, explorers, and privateers – a key figure in the defence of England during the summer of 1588. In 1596, Spanish forces captured Calais. This event provoked consternation in London – not only because Calais was among the most significant ports on the southern coast of the English Channel, but also due to its profound symbolic value: it had been the last English possession on the European mainland.[53] Among the troops who took part in the capture of Calais was the *Tercio de Stanley*. Definitive peace would not arrive until 1598. With Philip II gravely ill and eager to minimise the conflicts his successor would inherit, he initiated peace negotiations with Henry IV of France. These culminated in the signing of the Treaty of Vervins, which represented a kind of 'draw': Spain returned to France the strategic bases of Blavet and Calais. However, what might have been expected to bring relief to Elizabeth I was instead perceived as a betrayal. Despite England's substantial financial and military investment in the French theatre, the peace agreement merely left her without a critical ally in her ongoing war against Spain. Indeed, it is estimated that English intervention in France had cost £286,172 over four years. During that same period, military operations in the Low Countries incurred costs of between £500,000 and £600,000. These figures do not include the mounting expenses of privateering expeditions, which were becoming increasingly unsuccessful.[54] Moreover, since the outset of the conflict, English foreign trade had suffered severe setbacks, placing the national economy under considerable strain.[55] The resulting fiscal crisis forced the Crown to sell off royal lands and raise taxes to manage the spiralling debt.

The peace with France sparked discussions of peace within England as well. Queen Elizabeth's Privy Council became divided between those advocating for a dignified end to the prolonged war – led by William Cecil, Baron Burghley – and those determined to pursue hostilities at all costs, with Robert Devereux, Earl of Essex, at the forefront. England soon became saturated with pamphlets, many of them anonymous, defending both positions. The most notable among these were *Considerations of the Peace Now in Speech*, attributed to Burghley, and Essex's own *Apologie*.[56] The death of Burghley in August marked a pivotal shift in leadership, with his son Robert Cecil and the newly appointed Lord High Treasurer, Thomas Sackville, assuming his mantle.

Meanwhile, in Spain, the world was about to be shaken by a tragic event. Philip II, at the age of 71, died during the night of 12–13 September 1598 at the Royal Monastery of El Escorial – the very place he had chosen as his final resting site. His successor, Philip III, was a young man of 20 who lacked both the political acumen and the commanding presence of his father. Disinterested in matters of governance, the new monarch opted for a life of luxury, entrusting the reins of power to his inner circle. Chief among them was Francisco Gómez de Sandoval, Duke of Lerma, whom Philip III immediately appointed as his *valido*.[57] The Duke of Lerma had earned the prince's favour through flattery and opportunistic generosity; yet far from pursuing the common good, he was singularly focused on personal enrichment, inaugurating a period of maladministration that proved disastrous for Spain.

Returning to England, the Queen's Privy Council appointed the Earl of Essex as Lord Lieutenant of Ireland and sent him to lead an army in April 1599, replacing Sir Henry Bagenal, who had suffered a humiliating defeat at Yellow Ford at the hands of the Irish forces led by the Earl of Tyrone. Meanwhile, the network of spies maintained by Sir Robert Cecil across Spain reported that a large fleet was being assembled in various ports of Galicia. The anxiety and uncertainty, latent until then, erupted. They were facing the culmination of what Patrick Collinson famously dubbed the 'Nasty Nineties'.[58] Despite its efforts and continuous expeditions, the Royal Navy had been unable to destroy the Spanish fleet, which, in fact, was stronger at that time than it had been in 1588. Moreover, England's best troops were deployed in the Low Countries and Ireland, leaving the country virtually unprotected and without a standing army. The English had yet to grasp the temperament of the new Spanish king, Philip III, and began to fear that he might prove more aggressive than his father. Alarms were raised: a fleet was prepared for defence, and all available infantry was mobilised. Panic spread at court when, on 6 August, sentinels reported sighting a large Spanish fleet passing by the Isle of Wight. In reality, it was a group of Dutch merchant ships. The news plunged London into chaos and sparked civil disturbances. Eventually, it was discovered that no new Armada existed. This situation placed Sir Robert Cecil in a difficult position within the Privy Council. Although it is widely believed that the incident stemmed from an intelligence failure and the lingering fear of the Armada in England, historian Alan Williamson suspects that the Earl of Essex may have exploited the situation to destabilise the political standing of his rivals in government.[59] However, it did him little good. His campaign in Ireland was proving to be a complete failure. Rather than pursuing Tyrone's army in the north, he became entangled in battles and skirmishes in the south that lacked strategic value and merely depleted his resources and cost him men. Ultimately, when he resolved to attack the rebels' main bases in the north, one of his columns, under Sir Conyers Clifford, was ambushed in the Curlew Mountains, in what became known as the Battle of Curlew Pass, resulting in a decisive English defeat with hundreds of casualties, including Clifford himself. This led Essex to negotiate a peace with Tyrone without the Queen's consent, and he returned to England at the end of September. His arrival in London surprised everyone. Upon learning of the situation in Ireland, Queen Elizabeth was astonished, and Essex was summoned to appear before the Privy Council. His rivals, Cecil and Raleigh, seized the opportunity to accuse him of unauthorised negotiations and desertion. As a consequence, he was confined to his residence, Essex House,[60] which he soon began to fortify while gathering his supporters.[61] In October, Lord Mountjoy was sent to Ireland to continue the war, quickly achieving significant victories. In June 1600, Essex was tried and sentenced to permanent exclusion from court and stripped of his lucrative monopoly on sweet wines, effectively ruining him financially. Determined to overthrow the government, on 8 February he left his confinement accompanied by around 200 followers – including his stepfather, Sir Christopher Blunt, and Henry Wriothesley, Earl of Southampton – in an attempt to restore his former political standing. However, his enemies were already prepared. After a brief skirmish in which Blunt was wounded, Devereux was forced to retreat to Essex House, where he ultimately surrendered and was arrested. He and his supporters were tried for rebellion.[62] Twenty-six were fined; three received prison sentences; and seven were condemned to death. Only the Earl of Southampton was pardoned, his sentence commuted to life imprisonment (he would be released during the reign of James I). No clemency was granted to Captain Thomas Lea (executed on 14 February); Sir Charles Danvers; Sir Christopher Blunt (Essex's stepfather); Sir Gelly Merricke (also known as Sir Gelli Meyrick, Essex's steward); and Sir Henry Cuffe (regarded as the chief instigator of the conspiracy), all of whom were executed on 18 March. Lastly, Robert Devereux, Earl of Essex, was beheaded on 25 February.

The last major military operation of the war took place in Ireland in 1601. In context, since his landing, Lord Mountjoy had achieved significant advances against the rebels. Following his accession to the throne, Philip III considered the possibility of sending military support to Tyrone, which materialised in the form of a fleet of thirty-three ships under the command of Admiral Diego Brochero, with Pedro de Zubiaur as rear admiral, and a contingent of 4,432 men. The fleet departed from La Coruña on 2 September 1601. It comes as no surprise that, shortly before reaching Cork, their intended destination, a storm scattered the fleet. Most of the ships were forced to return to Spain, while a portion of the fleet under Brochero's command managed to land at Kinsale, where approximately 3,000 men of Juan del Águila's Tercio fortified themselves. English forces under Lord Mountjoy, numbering nearly 10,000 men, soon arrived and laid siege to the Spanish army. Despite the efforts of the Irish rebels to break the encirclement, by mid-January 1602, Juan del Águila presented his capitulation to Mountjoy. The English general offered generous terms: the Spanish Tercios were permitted to return to Spain, provided with ships and supplies for the journey, allowed to keep their banners and treasure, and Irish rebels who surrendered were granted amnesty. Thus, on 13 March, the troops of Juan del Águila arrived in La Coruña, having lost only about 100 men in the entire campaign. For his part, O'Neill remained in Ireland attempting to continue the resistance, but ultimately surrendered in March 1603 – just days after the Queen's death. The final years of her reign had not been easy, and as Christopher Haigh notes,[63] 'Elizabeth died unloved and almost unlamented, and it was partly her own fault . . . she ended her days as an irascible old woman, presiding over a war and failure abroad and poverty and factionalism at home'. If Philip had been devastated by the death of his daughter Catherine Michaela, Elizabeth experienced similar grief at the death of her friend and First Lady of the Bedchamber, Catherine Howard,[64] Countess of Nottingham, and even more so at the betrayal of Robert Devereux. She was also deeply affected by the death of her longtime and loyal advisor William Cecil. To this must be added a delicate political, economic, and military situation. And, not least, as described by Sir Robert Carey, she was burdened with remorse over the execution of her cousin, Mary Stuart. Now, however, England was entering a new era. The Tudor dynasty had come to an end, and the crown was to be placed upon another head. The chosen successor was James, King of Scotland. We shall speak of him later.

But beyond the major military operations previously mentioned, there were also lesser-known privateering expeditions and numerous naval battles that are worth summarising. Beginning with the privateering expeditions, despite some Pyrrhic victories, they had entered a period of decline. The war had led to the demise of the first generation of intrepid Elizabethan Sea Dogs, such as Drake, Hawkins, Frobisher, Devereux, Cavendish, and Greville. By the closing years of Elizabeth's reign, the new generation of privateers failed to match the brilliance of their predecessors. Hawkins Jr. was languishing in a Sevillian prison; Clifford, despite his tireless efforts, was ruined after twelve expeditions and forced to sell lands that had belonged to his family for generations. However, there were many others, less well known, who met with varied fates. Among them was Robert Dudley Jr., Duke of Northumberland, the illegitimate son of Robert Dudley, Earl of Leicester, who later legitimised him and named him heir. He took part in the capture of Cádiz in 1596, but his incursions into the Caribbean and his exploratory expeditions ended in failure. In 1605, he fled to Italy with his cousin and lover, Elizabeth Southwell,[65] whom he later married. They converted to Catholicism, and Dudley entered the service of the Grand Duke of Tuscany, Cosimo II, for whom he carried out significant engineering works. He died there in 1649. Another notable case is that of Christopher Newport, who, after a career as a privateer marked by some success but also serious defeats (including the loss of an arm in combat against Spanish galleons), became one of the founders of the first permanent English settlement at Jamestown, in the Chesapeake Bay (present-day Virginia, USA). He undertook several voyages to England to bring new settlers. On one of these journeys, he was shipwrecked and stranded on one of the Bahamian islands, where he survived for a year in a manner reminiscent of Robinson Crusoe. He died in 1617 on the island of Java, during a voyage in the service of the East India Company. We must also mention Anthony Sherley, who fought against the Spanish in the Low Countries and Normandy and took part in various expeditions to the Spanish Main. In 1598, following a voyage to Persia, he stopped in Spain on his way back to Europe, where he entered the service of Philip III, who eventually granted him command of a fleet. Finally, there is the case of Simon Bourman, who in 1601 attacked various towns in what is now northern Venezuela but was soon captured. He claimed to have been born in Spain and said that, as a child, he had been taken captive by the English and forced to convert to Anglicanism, which he later renounced, returning to Catholicism to avoid execution. In Cartagena de Indias, he provided the governor with intelligence about the routes used by English privateers and the weakest points in Spanish defences. This information proved particularly valuable in curbing the illegal salt trade, which was systematically exploited by English merchants and, even more so, by the Dutch.[66] Bourman was imprisoned in Spain until the end of the war. From that golden age, only the fate of Walter Raleigh remains to be recounted – one of Elizabeth's most prominent favourites. Until the queen's death, Raleigh alternated between serving in Parliament and holding the post of Governor of the Isle of Jersey. However, in July 1603, he was accused of involvement in the Main Plot[67] and subsequently imprisoned. In 1617, Raleigh was pardoned and granted permission to undertake a new expedition in search of El Dorado, accompanied by his son, Walter. However, the voyage ended in utter failure, culminating in the death of Walter Raleigh Jr. during a skirmish with Spanish forces. Upon his return, James I, under pressure from the Spanish Ambassador, the Count of Gondomar, and in accordance with the Treaty of London – a subject to be examined in greater detail in the following chapter – sentenced Raleigh to death for engaging in privateering in the Caribbean. He was executed by beheading in late October 1618, at the age of 65. And as we reflect on the fate of the leading figures of the so-called 'Golden Age', mention must be made of the Duke of Medina Sidonia. As previously discussed, his reputation remained unblemished following the failure of the 1588 Armada. Indeed, it was even strengthened in 1595, when he devised a system of relays between Andalusia and the principal ports of the viceroyalties of New Spain and Peru, using fast ships that kept both ends of the Atlantic in constant communication.[68] Nevertheless, the following year, the Cádiz raid did, in fact, threaten his prestige. However, he managed to restore his image in the eyes of Philip III thanks to his exceptional administration of the Indies fleets, which he equipped and outfitted annually with great diligence. He also efficiently oversaw the organisation of shipments of silver, gold, and other goods arriving at Andalusian ports from the Americas, which were subsequently handled by the *Casa de Contratación* in Seville. Equally significant was his role in organising a new fleet, known as the Armada of Flanders, composed of eight royal galleons, whose objective was to cleanse the Andalusian coast of privateers – primarily Dutch, but also English – who entered the Mediterranean under the pretext of trading with Italy. The Duke passed away in 1615.

Chapter 7

A RAINY DAY IN VALLADOLID: THE TREATY OF LONDON

From the mid-1590s, prominent political figures in England and Spain recognised the economic unsustainability of the war, and the necessity of considering a peace settlement. In the case of Spain, the war was threefold in nature, given its concurrent confrontations with France, the Netherlands, and England. The Armada of 1597 represented Philip II's final endeavour to oust Elizabeth from the throne. This subsequent failure led the Spanish monarch, aged 69 and in poor health, to the realisation that his life was drawing to a close and that the time for achieving a peace settlement had arrived. He was acutely aware that his heir, the future Philip III, lacked the intelligence, ability and even the will to rule that he himself or his beloved daughter Isabel Clara Eugenia had had. Consequently, the Spanish king granted his daughter sovereignty over the Netherlands, along with her husband, Archduke Albert. However, the stipulation was made that sovereignty would revert to the Spanish Crown in the event that the couple were rendered without issue.

Consequently, in May 1598, a mere four months prior to his demise, Philip II successfully concluded the Treaty of Vervins with France. However, Philip III was left with not only the world's largest empire, but also the daunting task of addressing a grave economic crisis and ongoing conflicts with England and the Low Countries. The young King, whose interests lay in a life of indulgence and extravagance, seemingly detached from the economic realities of his realm, placed his trust in the Duke of Lerma, whose actions were more focused on consolidating his own power and personal enrichment than on the general welfare of the nation. A notable example of this was the dissolution of the Junta de la Noche, a body that functioned as the Council of State under the reign of Philip II, and the subsequent dismissal of the former monarch's chief advisors. It was not until 1601 that Lerma became convinced that the time had come to seek peace with Elizabeth. A comparison of England's situation with that of Spain reveals certain parallels. The 1590s represented a particularly challenging period during the Elizabethan reign, characterised by a severe economic crisis triggered by consecutive years of poor harvests and the repercussions of the Spanish embargo on English textiles. This embargo was not compensated for by the booty captured by privateers, which merely enriched the shareholders who had invested in the chartering of the ships. Furthermore, it did not address the dynamic, albeit nascent, English trade with the Baltic countries and Russia. As Haigh, a historian who has been noted for his willingness to offer a critical perspective on Queen Elizabeth I, observes, in her later years, 'was shown to be politically bankrupt'[1] and she 'was not a wise and powerful statesperson, implementing the constructive policies she knew her nation needed: she was a nervous politician struggling for survival'.[2] This assessment is further supported by the following observation: 'Elizabeth died unloved and almost unlamented, and it was partly her own fault . . . she ended her days as an irascible old woman, presiding over a war and failure abroad and poverty and factionalism at home'.[3]

Furthermore, preoccupied with her own survival, she had failed to resolve the matter of her succession to the throne. However, England was fortunate in that the political acumen of Robert Cecil had already secured a clandestine accord with James of Scotland. Elizabeth's demise finally occurred on 24 March 1603. Concurrently, James was proclaimed King of England. This development was not met with disapproval by the Spanish monarchy, as the Scotsman was regarded as a 'friendly king'[4] and was considered the most favourable option, having relinquished his own candidacy for the throne some time earlier. Contacts were initiated for the purpose of resuming peace talks. It should be noted that this was not a return to peace talks in the traditional sense, but rather a resumption of diplomatic negotiations following the theatrical rapprochements between Elizabeth and the Duke of Parma prior to 1588, and a series of secret talks that had taken place in Boulogne in 1600, which had ultimately proved unsuccessful.[5]

This persuaded Elizabeth that a peace settlement in Spain was unfeasible, as she had minimal interest in the legacy she would leave to her successor. However, by June 1603, the political situation had undergone a radical transformation, with two Spanish envoys already in London to initiate peace negotiations. Philip III (Lerma, in fact) dispatched Juan de Tassis,[6] while Archduke Albert sent Charles de Ligne, both seasoned diplomats. The negotiations were relatively straightforward, with the primary common ground being the commitment of James to refrain from intervening in continental affairs, and of Philip to refrain from naming a Catholic king for England. Additionally, there was a guarantee of toleration of Catholicism on the island.[7]

Consequently, on 19 May 1604, the three-way negotiations formally commenced, with the foreign delegations being finalised: on the behalf of Spain, Juan Fernández de Velasco y Tovar, Duke of Frías and Constable of Castile, and Alessandro Robida, senator of the Duchy of Milan; and on the behalf of the Spanish Netherlands, Jean Richardot, president of the Privy Council, and Louis Vereyken, audencier of Brussels. The English side was represented by the following individuals: Robert Cecil, Earl of Salisbury and Secretary of State; Charles Blount, Earl of Devonshire; Thomas Sackville, Earl of Dorset and Lord Treasurer; Henry Howard, Earl of Northampton and Lord Keeper of the Ports; and Charles Howard, Earl of Nottingham and Lord High Admiral.[8] A total of eighteen sessions were convened at Somerset House (Westminster) between June and August. However, these proceedings were not without contention, as various English trade lobbies, representatives of the Dutch rebels and even the French ambassador, Christophe de Harlay, comte de Beaumont,[9] sought to thwart them. Ultimately, an accord was achieved on 28 August, which re-established the *status quo ante bellum*, that is to say, the pre-war situation. The thirty-six points of the treaty[10] can be summarised as follows:

- With regard to military aspects, the cessation of armed conflict between the two nations was a key element of the agreement, with Spain renouncing further support for the Irish rebels and England doing the same for the Dutch rebels. In the case of Flanders, as had been the case prior to 1585, English soldiers could be hired as mercenaries, but without the support of the English Crown.
- Furthermore, England committed to renouncing piracy and actively pursuing it.
- In the context of America, England acknowledged the Spanish monopoly on trade in the Americas, although it was stipulated that relatively free trade could be conducted 'in those places where it existed before the war' (clauses 20 and 24).

- The resumption of trade between England and Spain was permitted, provided that English Channel ports were used. Furthermore, English subjects were prohibited from trading Spanish goods with the Dutch rebels.
- In terms of religion, Spain renounced the restoration of Catholicism in England.

In other words, the treaty re-established the pre-1585 international political status quo, thereby recognising Spain's position as a hegemonic and all-encompassing empire. Furthermore, Philip III attained the objective of preventing English involvement in Spanish domestic affairs and of eradicating piracy, which had prompted his father to initiate the 1588 Armada. Consequently, the failure of the Armada had merely postponed Philip II's primary objectives by a period of 16 years. England (or Elizabeth I and her advisors), which had instigated hostilities with Spain to break the trade monopoly with America and claim its right to establish colonies, had failed. Which nation was strengthened by the signing of the treaty? A thorough examination of the treaty and the subsequent actions of the involved nations reveals the complexity of the treaty's interpretation by historians such as Geoffrey Parker, who asserts that 'James refused to satisfy the main objectives of Philip's war'.[11] Indeed, while in Valladolid, then the capital of Spain, the news was greeted with jubilation and joy, in London, although the negotiators claimed to have achieved an honourable diplomatic exit, it was generally regarded as a humiliating peace[12] which, in political terms, sent them back to the same situation of 1585 that had led them to provoke that long and costly war.

Months later, on 17 April 1605, Charles Howard, Earl of Nottingham, who had previously led the English ships that fought against the Spanish Armada of 1588 and had previously caused devastation to Cadiz in 1596, arrived in the port of Corunna.[13] On this occasion, however, his intentions were peaceful. He was accompanied by a retinue of more than 600 people and was en route to Valladolid to ratify the treaty that had been signed in London. The purpose of his mission was to officially attest to and be a witness to the signing of the document by Philip III. En route to the Spanish capital, Howard was greeted by a jubilant nation, which received him warmly and with elaborate festivities, such as those observed in Villafranca del Bierzo and Astorga.[14] The arrival of the English delegation in Valladolid coincided with the birth of the future heir to the throne, Philip IV. At that time, the Spanish court had moved on from the austerity of Philip II's reign and was engaged in a series of lavish festivities that left the

English in awe. On 26 May, Howard arrived at the gates of Valladolid.[15] The two most significant nobles of the court, the Duke of Lerma and the Marquis of Velada, proceeded to greet him. However, an untimely storm marred the welcoming ceremony, but over the next few days, Howard and his entourage were treated to numerous festivities, including the christening of the Prince of Asturias, culminating in the signing of the peace treaty on 9 June.[16] After the festivities, Howard left Valladolid on 18 June.[17] A peace that would last until 1625. But that is another story.

CONCLUSION

The failure of the Spanish Armada in 1588 has generated a vast historiographical output since the very moment Philip II's ships withdrew from the English Channel. From that point on, Elizabeth I's skilful propagandists sought to magnify a victory that has gone down in history for its (alleged) decisiveness and far-reaching consequences. This interpretation became firmly rooted in the collective imagination – not only in England – and has only begun to be critically reassessed in recent decades. Was the Battle of Gravelines truly a decisive engagement? Did it constitute a turning point in European history? For Mattingly, the defeat was indeed decisive, but only in terms of what might have occurred had the Armada achieved its objectives. Yet, from a historiographical perspective, this argument is problematic. When historians speak in the conditional, they are no longer dealing with history, but with counterfactual speculation – the infamous 'Cleopatra's nose' problem. A historian's task is not to analyse what could have happened, but to understand what actually did. Moreover, as demonstrated throughout this book, there are clear examples of naval battles during Philip II's reign that were genuinely decisive, such as Lepanto (1571) or São Miguel (1583).[1] As previously discussed, the 3.2 per cent loss of ships suffered by Medina Sidonia during his passage through the Channel does not justify concluding either English naval supremacy or that this was a decisive victory.

But what of its consequences? To begin with, as we have already noted, Elizabeth dispatched the so-called English Armada to Spain the following year, with disastrous results. In fact, the war dragged on until 1604. English historiography throughout the twentieth and twenty-first centuries has repeatedly claimed that Gravelines was a turning point in history – marking the decline of the Spanish Empire and the beginning of English maritime supremacy, allegedly rooted in superior ships, artillery, tactics, and seamen. Yet, as we have seen, throughout the Anglo-Spanish War of 1585–1604, England failed to achieve any further significant naval victories. Its successes were largely limited to raids on ports – most notably the sacking of Cádiz in 1596 – or the interception of isolated vessels by privateers, such as Drake's capture of the Portuguese *San Pedro*, Grenville's seizure of the Spanish *Nuestra Señora de la Concepción*, or the taking of the *Madre de Deus* by Frobisher and Clifford. However, a broader study of the subsequent decades – and even centuries – reveals that historical reality undermines this entrenched narrative. The defeat of the Armada did not, in fact, mark the beginning of the Spanish Empire's decline, which continued to expand. Under Philip II alone, Spanish territorial holdings in the Americas increased by 4.4 million square miles, nearly doubling the empire's size. In contrast, Elizabeth I's two attempts to establish permanent colonies – Roanoke Island and Saint John in Newfoundland – both ended in failure. Moreover, Spanish imperial growth persisted under Philip III and Philip IV. As for the navy, it did not lose its potency after 1588. On the contrary, as Mattingly himself concedes,[2] 'the defeat of the Armada was not so much the end as the beginning of the Spanish navy'. Philip II launched a reform that enabled Spain to maintain its dominance over the Atlantic route. In fact, throughout the entire sixteenth century, the Fleet of the Indies was lost on only two occasions. As Simpson rightly observes, 'the Spanish naval power had not been destroyed, and such defeats as had been suffered were due as much to the weather as to the English action . . . . For its part, the English navy, after the defeat of the Armada, had not succeeded in its two main goals – the destruction of the Spanish navy and the interception of the Spanish treasure fleets'.[3] Nor can it be argued that England emerged as a naval superpower in the immediate aftermath, since such a development only began to take shape a century later. In the early decades of the seventeenth century, Spain remained the hegemonic power. For instance, in 1625, the armies and navy of Philip IV achieved a decisive victory at Cádiz against an Anglo-Dutch fleet commanded by Willem van Nassau, Sir Edward Cecil, and Robert Devereux. They were decisively defeated by the Spanish forces under Manuel Pérez de Guzmán y Silva, Duke of Medina Sidonia.[4] The most significant example of Spain's enduring strength is perhaps the extraordinary expedition known as the Recapture of Bahia – at the time, the largest transatlantic deployment ever launched from Spain to the Americas. This campaign aimed to recover the Brazilian city of Bahia, which had fallen to Dutch forces the previous year. It is difficult to argue that a nation in decline could have mustered 52 ships and over 12,000 men, transported them across the Atlantic, and successfully defeated both the Dutch fleet and army. In reality, it was not until the 1640s that Spain entered a period of serious crisis, eventually leading to the definitive loss of its Dutch territories and the separation of Portugal. First the Dutch Republic, and later Louis XIV's France, assumed the mantle of European hegemony. Nevertheless, the naval strength of seventeenth- and eighteenth-century Spain should not be underestimated. The Spanish navy continued to produce remarkable sailors, scientists, and naval engineers whose contributions to maritime innovation were considerable, though their names are likely unfamiliar to most English – and even Spanish – audiences. Figures such as Antonio de Gaztañeta e Iturribalzaga (1656–1728), the brilliant Jorge Juan (1713–73), and Admiral Cosme Damián Churruca (1761–1805) stand out among them. One must also acknowledge iconic vessels such as the *Santísima Trinidad* (1769), arguably the most formidable ship of the line ever constructed – a technical feat inconceivable in a truly declining nation. Even the loss of the Azores in 1640 did not collapse the Spanish Empire; by then, Spain had already developed faster, more capable ships that no longer depended on that strategic mid-Atlantic stopover. If, as it is often claimed, the failure of the 1588 Armada marked the beginning of Spain's decline, we might paraphrase Will Durant's[5] famous remark about Rome and say that the fall of the Spanish Empire 'was not an event but a process spread over 200 years. Some nations have not lasted as long as Spain fell.'

So where does the idea that Gravelines was a decisive victory and a much-repeated turning point originate? It goes without saying that the failure of the Armada was celebrated throughout England. However, the subsequent development of Elizabeth's reign gradually tempered the exuberance of the summer of 1588. Not only because military affairs seldom brought good news from the port of Plymouth, but also because the English economy suffered due to a series of poor harvests during the 1590s. As Haigh notes, 'when the economic and military circumstances of the post-Armada years proved so difficult, the old image was tarnished, and the old promises were shown to be hollow'.[6] Moreover, 'in the new and bitter world of the 1590s, Elizabeth was shown to be politically bankrupt'.[7] The end of the Tudor dynasty appears to have buried the memory of the victory, as this supposedly momentous event was not commemorated on either its first or second centenary. Remarkably, England's most prominent author at the time, William Shakespeare, did not write any plays about his queen, despite the episodes of fervent nationalism and triumphalist sentiment that arose during her reign. In fact, some scholars suspect that, if he did so, it was in a subversive manner.[8] Interest in Elizabeth I and the Armada of 1588 was not revived until the nineteenth century. In this context, Douglas Knerr provides a compelling analysis of the historiographical resurgence, tracing the first major studies of the battle to the 1830s, within the framework of what has come to be known as the Whig interpretation[9] of history – marked by historical romanticism and nationalist zeal. Among the most representative figures of this tradition were James A. Froude and Thomas Carlyle. As Knerr continues, 'the Whig view became a Victorian era standard and left an indelible mark on Armada historiography. . . . The Whig historians exploited the story of the Armada on many pages of colourful, descriptive, if often biased historical accounts.' It is true, however, that in Froude's work, Elizabeth's role is criticised, and the credit for the victory is instead attributed to Burghley and Walsingham. With the turn of the century, these myths took on a more scientistic tone, particularly following Corbett's *Drake and the Tudor Navy* (1898), which argued that 'the English fleet was on a fair equality with the Armada in the galleon class and possessed an overwhelming preponderance of gunpowder' – a work that also lends credibility to the infamous bowls anecdote. This glorification of Drake's role was further echoed in J.A. Williamson's *The Age of Drake* (1938). However, this line of analysis – which became the model for works on the subject throughout the twentieth and twenty-first centuries – fails in that it decontextualises the battle, building hypotheses that are unsustainable when confronted with the subsequent course of the war. Historiography has merely lent a veneer of legitimacy to a narrative that has shaped the popular understanding of the failed Enterprise of England. Yet historiography alone does not create a 'discourse' or a cultural framework capable of producing a particular worldview. Especially prior to the late twentieth century, historical interest remained largely confined to academic institutions. In the nineteenth century (and arguably even earlier), it was politics and the arts that shaped this worldview. It is no coincidence that this century witnessed the convergence of various factors that would create and consolidate the myths surrounding the Armada and Elizabeth – a figure who would captivate Romantic artists. And it is hardly surprising that a woman with such a biography inspired admiration: motherless from a young age, imprisoned by her half-sister, a champion of Anglicanism against the Catholicism of the mighty King of Spain; a virgin queen, yet surrounded by numerous suitors and a court teeming with secret romances. In contrast, her great adversary, Philip II of Spain, has been portrayed as a heartless Catholic fanatic. It is not the purpose of this book to delve deeply into this subject, but it is illustrative to turn to opera to juxtapose the figure of the Queen, as depicted in Donizetti's so-called *Tudor Trilogy*, with the image of the Spanish king in Verdi's *Don Carlo*.[10] Indeed, one of

Romanticism's legacies has been what we might call a certain re-creation – or idealisation – of the past. A telling example of this phenomenon is the invention of traditions, as brilliantly analysed by E. H. Thompson. Yet the nineteenth century was not merely the period in which many traditions still familiar to us were created; it was also the era in which the foundational myths of nations were widely disseminated. These myths served to foster a sense of unity and national identity, reinforcing traits or values that each nation typically takes pride in, and linking present rulers to those of the past in a legitimising continuity. Above all, they exist to legitimise narratives. In Spain, examples of such foundational myths include the Battle of Covadonga (722) and, alternatively, the so-called Reconquista (a term increasingly subject to revision) and the Battle of Las Navas de Tolosa (1212). But such myths can be found in every country. Since the medieval period, King Arthur[11] has functioned as the foundational myth of the English nation, but the British Empire would require its own myth of origin. It is hardly coincidental, then, that the myth of the Armada was constructed and amplified at the very height of imperial British power – and, curiously enough, at the time of the tricentennial of the Enterprise of England and, notably, at the moment when another woman was being crowned: Queen Victoria. The symbolic connection legitimised the present through the past. Thus, although historiography remained under the influence of Froude's interpretations, a 'frame' began to emerge – in the sense described by George Lakoff[12] – in which the reign of Elizabeth I and her victory over the Armada became central to the narrative explaining the success of Victorian Britain. During the tricentenary, official propaganda fully embraced the myth. It offered a brilliant echo of Elizabethan propaganda in the service of British *ethos*. Commemorative medals were struck, books were published, monuments were erected, engravings widely circulated. It mattered little, as Julia Walker notes, that Queen Victoria 'vocally asserted that she shared little in common with her early modern predecessor',[13] given that the 'Grandmother of Europe' – a woman who understood her royal duty as including the assurance of dynastic continuity, and who therefore likely harboured some disdain for the 'Virgin Queen' – nonetheless recognised that the true goal was to legitimise an empire and reinforce a national narrative. That narrative, that *ethos*, defined how the British are 'truly' supposed to be – not necessarily how they are, but how they like to be perceived: brave, composed, charmingly arrogant. What better example of this than the story of Sir Francis Drake insisting on finishing his game of bowls after the Armada had been sighted? 'There's time for that [playing bowls] and to beat the Spaniards after.'[14] Of course, this episode never actually happened – but that is beside the point. Its significance lies not in its historical accuracy, but in how it encapsulates the self-image that Britons wish to project. A similar case is the famous Tilbury speech. It has already been historiographically demonstrated that it did not take place on 8 August 1588, but rather several days later, when the Spanish threat was already steering toward the Irish coast. But again, this does not seem to matter. In a film as recent as *Elizabeth: The Golden Age* (2007), the myth is still reinforced, with the speech portrayed as occurring just before the Battle of Gravelines. The film's screenwriters, William Nicholson and Michael Hirst – both British, both educated at prestigious institutions such as Cambridge and Oxford – are among the UK's most successful writers. Is it possible they did not know the actual date of the speech? It doesn't matter. The point was not to recount what truly happened (for that, buy a good history book). The goal was to craft a dramatic momentum and, through myth, to reinforce a narrative: that of a queen – and a nation – strong, independent, indomitable. It is about reinforcing discourse through correspondences drawn from foundational myths.

Moreover, the aim was never to explain to the world the origins of the British Empire, but rather to justify the causes of its unparalleled dominance in history: the reasons for English superiority. A superiority that viewed the Spanish as a backward and decadent nation – a perception that may have been accurate in the nineteenth century, but certainly not during the sixteenth, seventeenth, or even the eighteenth centuries. Yet the nineteenth, twentieth, and twenty-first centuries have been dominated by what we might call the 'Anglosphere'. Initially led by Great Britain, its cultural supremacy would later pass to the United States, more precisely to those who were identified at the end of the last century as WASPs (White Anglo-Saxon Protestants), who, significantly, look to England as their foundational myth. This explains the repetition of historical clichés. While English sailors and scientists are remembered as the pioneers who advanced the world, the Spanish are depicted merely as Catholic fanatics – incapable of invention, uninterested in scientific practice, and motivated only by the desire to plunder the gold and silver of the Indigenous peoples of the Americas. Such narratives, for example, completely silence figures like Domingo de Soto (1494–1560), who anticipated much of the laws of motion and universal gravitation later formulated by Newton (and first explored by Galileo); or Félix de Azara y Perera (1742–1821), a military officer and scientist whose ideas Darwin 'borrowed' in the development of his theory of species. In the twentieth century, popular culture has undoubtedly been dominated by cinema, which has become the primary vehicle for the dissemination of narratives. Film has done little more than reinforce the myths first constructed during Romanticism or bolster the ideals of Anglo-Saxon *ethos*. One example suffices: in *Mutiny on the Bounty* (1962), a character states, 'It's funny that nobody's ever heard of this before, then. No one ever heard of the potato until Sir Francis Drake brought it from South America. It altered the European economy.' This statement is pure fiction. The potato was introduced to Europe in the mid-sixteenth century by the Spanish soldier, chronicler, and explorer Pedro Cieza de León, who brought the first potato plants to Europe and described them in several chapters of his *Crónica del Perú* (1553). It is now believed that the potato reached the British Isles through English merchants based on the Andalusian coast. Film (and, more recently, television series) has also played a central role in forging the image of Elizabeth I as a brave, independent, intelligent and daring queen. Without her representation in fiction, it would be difficult to consider the daughter of Henry VIII one of the greatest British monarchs – especially given the limited objective achievements of her reign. Early cinematic portrayals of Elizabeth focused more on romantic affairs than political matters. However, the rise of Hitler (as in *Fire Over England*, 1937) and the outbreak of the Second World War (*The Sea Hawk*, 1940) gave rise to the myth of a Queen confronting an implacable enemy. In contrast, Philip II was cast as a sixteenth-century Hitler. The Anglosphere has judged both monarchs with remarkable bias. Beyond the wartime propaganda films of the mid-twentieth century, Philip II remains entrenched in the Anglo-Saxon imaginary as a religious fanatic. But were not all monarchs of his time fervently religious? Was he any more fanatical than Elizabeth, who established the Court of High Commission, under which hundreds of English Catholics were persecuted? In the British collective imagination, Elizabeth I is celebrated as one of the greatest monarchs in history. Yet her greatest 'achievement' was provoking Philip II. She initiated a war she could not win and, contrary to popular belief, did not found an empire. Her attempts to colonise the Americas ended in failure; she did not revolutionise naval warfare, nor did the Royal Navy achieve maritime dominance under her rule. As we have shown, Spain continued to dominate the Atlantic well into the seventeenth century, until it was eventually surpassed by the Dutch Republic. Elizabeth I had one overriding goal: to remain on the throne. For this, she tolerated the rampant factionalism of her reign and the repeated insubordination of her privateers, who, as we have seen throughout this book, pursued personal gain rather than serving the

Crown (with very few exceptions). The evidence is so overwhelming that some historians have concluded that her mere survival was a tremendous achievement. If Elizabeth's survival is deemed a major success, then it is equally justifiable to consider Philip II's ability not only to preserve but to expand his empire – despite being besieged by Ottomans, Portuguese, French, English and Flemish opponents – as a monumental feat. The same can be said of his successors: despite the supposed Spanish decline, during the seventeenth and eighteenth centuries Spain lost only a few thousand miles of territory – out of an empire that once spanned 20 million square miles. As for Philip II's portrayal in film and television, it remains entirely divorced from historical reality. The distortion reaches a climax in the recent adaptation of *The Serpent Queen*, where a young Prince Philip is portrayed as a psychopathic, violent and cruel figure – despite the fact that, at that age, he had already served as regent of Spain during his father Charles V's absence, demonstrating a maturity and political acumen that astonished both the Castilian nobles of the Royal Council and foreign ambassadors. Not to mention the persistent misrepresentation of his appearance, often depicted as North African in complexion, whereas he was in fact fair-skinned, blond and blue-eyed. But, of course, all this forms part of a narrative construction that the reader will by now have recognised.

At the beginning of this book, we posed the question: Was the Battle of Gravelines the Tudors' Greatest Naval Battle? In light of the evidence presented, we must answer affirmatively. It is true that the English Armada of 1589 was numerically superior, but we have also seen how Drake failed utterly to capitalise on it. More significantly, we can assert that Gravelines stands as one of the most important naval battles in British history – not for its result, nor for its consequences, but for its role in the collective imagination: first English, then Anglo-American. Once again, we must return to cinema to understand why. I cannot conclude without referencing John Ford, a filmmaker of boundless insight, who put it best in *The Man Who Shot Liberty Valance*, when the journalist Maxwell Scott says to Ransom Stoddard: 'This is the West, sir. When the legend becomes fact, print the legend'. The Spanish Armada is the foundational myth of the British Empire. To paraphrase: This is the Tudors' England.

POSTSCRIPT: THE IRONY

Let me move to Portsmouth on 21 May 1662 to finish the book. There, church bells were ringing with joy. King Charles II had just married a Portuguese princess, Catherine of Braganza. This sealed an alliance between the two countries that was to last for centuries. The King's joy could not have been greater, as he also received a substantial dowry and privileges to trade with the Portuguese colonies. However, this wedding held a secret that concerned the newly crowned Queen of England. Her mother, Queen of Portugal through her marriage to John IV of Braganca, was Luisa Maria Francisca de Guzman y Sandoval, daughter of Manuel Alonso Perez de Guzman, VIII Duke of Medina Sidonia, the first-born son of Alonso Perez de Guzman, VII Duke of Medina Sidonia. What a twist of fate! The man who had commanded the Armada of 1588 in an attempt to depose Elizabeth I. His Armada had failed, but now his great-granddaughter was crowned Queen of England. Ironies of history.

NOTES

Chapter 1: Adam's Will and Testament

1   Sevilla González (2016). It was not the first diplomatic treaty signed by the kings of England and Spain. Relations can be traced back to the year 1170, with the marriage agreement between Alfonso VIII of Castile and Eleanor Plantagenet, daughter of Henry II of England and Eleanor of Aquitaine, a wedding that would take place seven years later (Vigil Montes [2021]). There are other treaties that turned English princesses into Spanish queens, such as the case of the intelligent and influential Catherine of Lancaster (queen between 1390 and 1406), wife of Henry III of Castile, or Eleanor of Castile (queen between 1279 and 1290), first wife of Edward I of England. Diplomatic agreements were also reached without marriage alliances, such as the one signed between Henry III of England and Alfonso X of Castile in 1254, for instance.

2   Daughter of Edward IV, sister of Edward V, and niece of Richard III – the last three kings of the House of York.

3   With all the nuances we may wish to include, it could be described as a kind of Commonwealth *avant la lettre*: the Crown was shared and there was a willingness to cooperate, yet both kingdoms fully retained their sovereignty with respect to one another. I may add that, while the Crown of Castile was composed of various kingdoms but with power centralised in Castile, the Crown of Aragon had an equally confederal structure in which power was initially centralised in Catalonia (until its crisis in the fourteenth century), and subsequently in the Kingdom of Valencia.

4   Mesa Gallego (2021).

5   An independent kingdom that posed a strategic threat, as it was an ally of France south of the Pyrenees.

6   Great-grandfather of Thomas Howard, 1st Earl of Suffolk, who will later emerge as one of the main protagonists in our account.

7   During the reign of Edward VI, Herbert allied himself with the ambitious Northumberland, even arranging the marriage of his eldest son, Henry, to Lady Catherine Grey, the sister of Lady Jane Grey – whom Northumberland intended to place on the English throne

following the death of the weak and ailing king. However, when Herbert realised that Elizabeth was likely to be proclaimed queen, he promptly expelled Lady Catherine Grey from his household and annulled the marriage contracted with his son. In fact, Lady Catherine remained a potential heir to the English throne, and it appears that Philip II even considered marrying her to his own son, Charles of Castile. Nevertheless, the idea ultimately came to nothing.

8   His father had already inflicted a serious facial injury on Francis I in a previous accident. One had to be cautious with this family. Among their descendants was the legendary General Bernard Law Montgomery.

Chapter 2: The Swell Before The Storm

1   Philip II had prohibited slavery in the New World. However, in the Caribbean islands, there was a desperate need for labour. The local populations had been nearly annihilated by conquest, wars and epidemics. As a result, plantation owners purchased African slaves by the hundreds, while local authorities turned a blind eye to the practice. This trade proved to be an extremely lucrative business for both parties.

2   Martínez Ruiz (2022), p. 403.

3   The fate of these sailors varied greatly. It appears that those who were trapped in San Juan de Ulúa were eventually released, either entering the service of Spanish nobles in the region or establishing themselves as free individuals. A small group joined the crew of a French pirate ship and returned to England. However, another group, which had disembarked in Panuco (modern-day Tampico, Mexico), met a far worse fate. Attacked by Chichimeca natives, the seventy-seven survivors were detained by the Inquisition and accused of being 'Protestant heretics'. Those under the age of 16 were sent to convents to be educated in the 'true faith'. The remaining sixty-eight were sentenced to receive 200 lashes and to row in galleys for periods ranging from 4 to 12 years. One of them, George Ribley, was selected to be executed. The *auto-de-fe took* place on 28 February 1574, during which a group of French corsairs was also whipped. Ribley and the French corsair Martin Cornú were executed by being burned at the stake. Alberro (2015) and Ita Rubio (2017), p. 27 and pp. 42–8.

4   Hawkins (in Burrage, 1906), Casas (2015).

5   Fray Pedro Simón, pp. 88–93.

6   The Maroons (*cimarrones* in Spanish) were fugitive black slaves who managed to establish settlements in the territories now occupied by Panama and the Colombian coast. Their economy was primarily based on subsistence agriculture, supplemented by robbery of travellers, attacks on plantations, and occasional collaboration with corsairs and pirates. The Spanish authorities launched numerous military expeditions to destroy these settlements, albeit with little success. The fruitful collaboration between Drake and the Maroons forced the

Spanish to shift their policy from harsh repression to negotiation. Thus, in 1579, an agreement was reached with a large group of Maroons, allowing them to settle in the town of Santiago del Príncipe (near Nombre de Dios). In exchange for autonomy, land and freedom, they pledged to help in the defence of Panama. The town was placed under the leadership of a Maroon leader who called himself King Don Luis de Mozambique. From that point onward, the Maroons became loyal allies of the Spanish Crown. (Laviña, et alt., 2015, and Segas, 2017).

7   Amaya Palacios (2021), p. 17. The total loot consisted of 190 mules, each carrying 300 pounds of silver, amounting to a total of 57,000 pounds. The exact distribution of such a substantial treasure remains unknown, as it would have been divided among the English, French, and the Maroons. Therefore, any figures cited regarding the division are purely speculative.

8   As an extension of the previous note, the exact amount of silver Drake ultimately managed to take remains unknown. In any case, it was an extraordinary fortune, despite Lucena Salmoral – always so sparing in his praise for Drake – claiming that it was a modest haul for such great effort (Lucena Salmoral, 1992, p. 45). As for the silver that Drake buried, it has sparked the imagination of generations of treasure hunters for centuries, yet no trace of it has ever been found. Sources indicate that the Spanish authorities most likely succeeded in recovering it. Regarding Guillaume Le Testu, who was wounded during the attack, he was captured by Spanish troops and executed in Nombre de Dios.

9   It was the third circumnavigation of the globe after the expedition of Magellan and Elcano (completed in 1522) and the less well-known, yet arguably more epic, expedition of García Jofre de Loaisa and Andrés de Urdaneta (completed in 1536).

10   Fray Pedro Simón, p. 92.

11   At the current exchange rate, approximately £51 million. Source: https://www.nationalarchives.gov.uk/currency-converter/#currency-result

12   *The voyage of Master Andrew Barker of Bristol, with two ships, the one called the Ragged staffe, the other the Beare, to the coast of Terra firma, and the Bay of Honduras in the West Indies, in the yeere 1576,* in Hakluyt (1907). In: https://www.perseus.tufts.edu/hopper/text?doc=Perseus%3Atext%3A1999.03.0070%3Anarrative%3D740. Accessed 23 May 2024.

13   Edwards (1992).

14   Charter to Sir Walter Raleigh. In: https://avalon.law.yale.edu/16th_century/raleigh.asp. Accessed 24 May 2024.

15   *The first voyage made to the coasts of America, with two barks, where in were Captaines M. Philip Amadas, and M. Arthur Barlowe, who discovered part of the Countrey now called Virginia, Anno 1584. Written by one of the said Captaines, and sent to Sir Walter Ralegh*

knight, at whose charge and direction, the said voyage was set forth. In Hakluyt, op. cit. In: http://www.perseus.tufts.edu/hopper/text?doc=Perseus%3Atext%3A1999.03.0070%3Anarrative%3D670. Accesses 26 May 2024.

16   *The voiage made by Sir Richard Greenvile, for Sir Walter Ralegh, to Virginia, in the yeere 1585*, in Hakluyt, op. cit., in: http://www.perseus.tufts.edu/hopper/text?doc=Perseus%3Atext%3A1999.03.0070%3Anarrative%3D671. Accessed 26 May 2024.

17   A Letter from M. John White to M. Richard Hakluyt written in February 1593, in Hakluyt, op. cit. In: http://www.perseus.tufts.edu/hopper/text?doc=Perseus%3Atext%3A1999.03.0070%3Anarrative%3D704. Accessed 27 May 2024.

18   About this subject, Sgroi (2003).

19   Sgroi (2003), p. 7.

20   Ménard, op. cit., p. 265 y Cell (1969), p. 178.

21   As the reader will know, she was proclaimed Queen in November 1558, but her coronation was in January of the following year.

22   The Sultan's decision to conquer Malta was probably influenced by the raid in 1564 on a convoy of twenty Ottoman ships by seven Maltese ships under the command of Mathurin Romegas, an admiral of the Order of St John and a fast privateer. As a result, a ship belonging to one of his chief advisers was captured, and among its many illustrious passengers was the Governor of Egypt, in addition to booty worth around 80,000 ducats (about £5.5 million). Balbi (2005), p. 29.

23   It is important to note that he had previously attempted to proceed overland, but his advance was halted at the Siege of Vienna in 1529, where he encountered a coalition of Christian armies.

24   The Order was suppressed by Henry VIII in 1540, leading to the execution or imprisonment of some of its principal members. Queen Mary I restored it in 1557, but the number of new knights admitted was minimal. Consequently, English knights were often grouped together with the German knights, another Langue of the Order that was experiencing a decline. In 1559, Queen Elizabeth once again prohibited the Order.

25   The previous Holy League, formed in 1560 to conquer the island of Djerba (modern-day Tunisia), was a complete disaster due to the lack of coordination among the three fleets involved – Papal, Venetian and Spanish.

26   An intriguing figure. Born in Calabria, he was captured by the Ottomans and converted to Islam. His innate talent for naval warfare ultimately elevated him to the rank of one of the greatest admirals of the Sublime Porte.

27   Woodehead (2009).

28   Ibid.

29   Castries (1918), pp. XI–XII.

30   MacLean, Matar (2011), p. 52.

31   There were two other candidates who could have claimed the throne but, due to their alliance with Philip of Spain, they renounced their claims. These were Emmanuel Philibert, Duke of Savoy, and Ranuccio Farnese, an 11-year-old and the eldest son of the Duke of Parma.

32   His mother was a low-class commoner named Violante Gomes, though she was known as '*Pelicana*'.

33   Letter from Wotton to Walsingham. Calendar of State Papers Foreign: Elizabeth, Volume 14, 1579-1580. In: https://www.british-history.ac.uk/cal-state-papers/foreign/vol14/pp43-52. Accessed 21 July 2024.

34   Document 402. In: https://www.british-history.ac.uk/cal-state-papers/foreign/vol14/pp388-406. Accesses 22 July 2024. This request will be repeated in May 1580, document 206, in: https://www.british-history.ac.uk/cal-state-papers/foreign/vol15/pp181-194. Accessed 22 July 2024.

35   Document 410. In: https://www.british-history.ac.uk/cal-state-papers/foreign/vol14/pp388-406. Accessed 22 July 2024.

36   Document 437. The report from Burghley's office on the battle. Dated 30 September 1580. In: https://www.british-history.ac.uk/cal-state-papers/foreign/vol14/pp419-436. Accessed 22 July 2024.

37   An example of this can be found in document 179 of the Calendar of State Papers Foreign: Elizabeth, Volume 15, 1581–82, dated 10 May, in which Don Antonio grants Elizabeth the gold and ships stored in the fortress of Saint George in Amina (Ethiopia) and issues a letter of marque in case they encounter any Spanish vessels on their return journey. In: https://www.british-history.ac.uk/cal-state-papers/foreign/vol15/pp154-167. Accessed 22 July 2024.

38   Quinn (1979), p. 210.

39   Document 182 dated 12 May, Letter from Cobham to Walsingham. In: https://www.british-history.ac.uk/cal-state-papers/foreign/vol15/pp167-181. Accessed 22 July 2024.

40   Filippo de Piero Strozzi (1541–82) was an Italian *condottiero* in the service of the French Crown. Although his mother was Laudomia de' Medici, she came from a minor branch of the powerful Florentine family, and consequently, Strozzi was not related to the Queen Mother, Catherine de' Medici. Nevertheless, Catherine always showed great affection for the young and effective general.

41   Document 530, Letter from Diego Botelho to the Queen, dated 1 February 1582. In: https://www.british-history.ac.uk/cal-state-papers/foreign/vol15/pp478-491. Accessed 22 July 2024.

42   Quinn, op. cit., p. 212. James (2012), p. 12.

43   For the reader who wishes to gain an in-depth understanding of this pivotal battle, the following book (available only in Spanish) is indispensable: Antonio Luis Gómez Beltrán, *Islas Terceiras. La Batalla Naval de San Miguel*. Ediciones Platea, 2017. For this chapter, I have also relied on Rodríguez Garat (2023).

44   Roncière (1899), p. 192.

45   Ibid.

46   Jacques-Auguste de Thou: *Histoire universelle*. Volume 8, page 596. In: https://archive.org/details/histoireuniverse08thou/page/596/mode/2up?q=madeira. Accessed 25 July 2024.

47   Fréné (2015), Jones (2022).

48   James (2012), p. 3.

49   Document 112 dated 24 June 1583. In: https://www.british-history.ac.uk/cal-state-papers/foreign/vol16/pp98-109. Accessed 25 July 2024.

50   Document 176 dated 19 October 1583. In: https://www.british-history.ac.uk/cal-state-papers/foreign/vol18/pp144-154. Accessed 25 July 2024.

51   Roncière, op. cit., p. 205.

52   Document 393 dated 30 January 1584. In: https://www.british-history.ac.uk/cal-state-papers/foreign/vol18/pp319-334. Accessed 25 July 2024.

53   Document 408 dated 7 February 1854. In: https://www.british-history.ac.uk/cal-state-papers/foreign/vol18/pp335-346. Accessed 25 July 2024.

54   His nephew Richard Shelley died in prison in 1586 for practising the Catholic faith.

55   https://doi.org/10.1093/ref:odnb/23634. Accessed on 1 June 2024.

56   García Hernán (1999), p. 149. Citing the documents: ACIS. E. 823, 150–158, minutes of the Council dated 7 July 1571, which approved Ridolfi's plan; and from the Archive of the House of Alba, Box 7, fol. 5EJ: 14 July 1571, in which the king ordered the Duke of Alba to proceed with the invasion.

57   Nicholas Throckmorton fell from favour after the Ridolfi Plot, as he was suspected of being involved. He was imprisoned but released without trial; however, he never regained the Queen's trust.

58   Margaret of Parma (1522–86). The illegitimate daughter of Charles I of Spain and, consequently, the half-sister of Philip II of Spain, she served as the governor of the Seventeen Provinces from 1559 to 1567. She married Ottavio Farnese, Duke of Parma, and was the mother of Alessandro Farnese.

59   They were Fernando Álvarez de Toledo, 3rd Duke of Alba (1567–73),
     Lluís de Requesens (1573–6), John of Austria (1576–8), and Alessandro
     Farnese, Duke of Parma (1578–92).

60   He is also known as John Norreys or Norrys. In this book, we will
     consistently use the form 'Norris' to refer to him.

61   Miller (2013).

62   Gallegos (2013), p. 217.

63   Philip II had declared William of Orange a traitor and offered a reward
     of 25,000 ducats (equivalent to approximately £2.3 million or $2.9
     million today) for his head. Since Gérard was executed, the money was
     awarded to his parents.

64   Nexon (2009), p. 223

Chapter 3: 1585–1588: Early Battles

1   Carleill (1551–93) was an exceptional military leader who served
     in Flanders under the command of William of Orange, where he
     distinguished himself with his victory during the Siege of Middelburg
     (1572–4) and later in Ireland. By 1588, he held the position of Governor
     of Ulster and was one of the few Englishmen who treated the survivors
     of the Spanish Armada with compassion.

2   Ortigueira Amor, Poggio Capote, et alt. (2014).

3   González López (1970), p. 289.

4   In 1586, during his return voyage to Spain, Pedro Sarmiento de
     Gamboa's ship was seized by three vessels under the command
     of Walter Raleigh and taken to England, where he was eventually
     released. However, as he travelled back to Spain, near Paris, he was
     captured by a group of Huguenots who demanded a ransom from
     Philip II.

5   Regarding this expedition, Ita (2017) provides an account. As for the
     fate of the Spanish sailor Tomé Hernández, sources differ in their
     accounts. According to Ita, he managed to escape along the Peruvian
     coast and later returned to Spain. In contrast, González Ochoa states
     that he arrived in Plymouth with the survivors of the expedition and,
     after spending several months in prison, was granted permission to
     return to Spain.

6   To recount this incursion, I will draw upon documents from Spanish
     archives as well as two anonymous reports written by Italian merchants
     who witnessed the attack. These reports are preserved in the *Archivum
     Romanum Societatis Iesu* (hereafter referred to as Document A) and in the
     Archivio di Stato in Florence (hereafter Document B), both of which were
     recovered by Tanturri (2012). In addition, I will rely on the analysis
     provided by Gómez Beltrán (2022).

7   Letter dated on 13 May 1587. Paris Archives, K. 1448. 118. It can also be read at the Calendar of State Papers, Spain (Simancas), Volume 4, 1587-1603. In: https://www.british-history.ac.uk/cal-state-papers/simancas/vol4/pp78-92. Accessed 19 September 2024.

8   It is not possible to determine the exact number of ships involved, beyond the twenty-three that departed from Plymouth on 12 April. Corbett (p. xx, 1898) provides the following list: from the Queen: *Elizabeth Bonaventure* (550 tons, Sir Francis Drake, flagship), *Golden Lion* (550 tons, William Borough, Vice-Admiral), *Dreadnought* (400 tons, Thomas Fenner), *Rainbow* (500 tons, Henry Bellingham), and two pinnaces: *Spy* (50 tons, Clifford), and *Makeshift* (50 tons, Bostocke); from the Lord Admiral: *White Lion* (150 tons), and one pinnace: *Cygnet* (25 tons); from the Levant Company's Squadron: *Merchant Royal* (400 tons, Flick, Rear-Admiral), *Susan* (350 tons), *Edward Bonaventure* (300 tons), *Margaret and John* (210 tons), *Solomon* (200 tons), *George Bonaventure* (150 tons), and *Thomas Bonaventure* (150 tons); from Drake: *Minion* (200 tons), *Thomas* (200 tons), *Bark Hawkyns* (130 tons), and *Elizabeth* (70 tons); other vessels: *The Little John* (100 tons); and three pinnaces: *Drake* (80 tons), *Speedwell* (50 tons), *Post* (30 tons). Total: sixteen ships and seven pinnaces. In any case – and this was common practice – Drake had been instructed to invite any English ships he encountered in the English Channel to join his expedition. This practice often resulted in the 'dragging along' of various pirate vessels into his ventures.

9   According to Document B, other sources report that the ships were flying French and Flemish flags (Calvar Gross et al. 1988-1993, Vol. III, Tomo II: 635. Doc. 2205. Archive: MN, Ms-496, Col. FN, t.XXX, doc. 277; proc. CDA).

10   Acuña was temporarily in command of the nine galleys of the Spanish Galley Squadron, whose Captain General, Martín de Padilla, was in Málaga at the time.

11   Corbett (1898), p. 24.

12   Tanturri (2012), p. 74.

13   Ridella et al. (2016), p. 13.

14   Given that the Republic of Genoa and England had signed a peace treaty, Drake justified in his report that the sunken ship originated from Ragusa, an ally of Spain.

15   State Papers, Domestic, CC.46.d1646.

16   https://www.nationalarchives.gov.uk/currency-converter/#currency-result. Accessed 22 September 2024.

17   Hutchinson (2013), p. 203.

18   https://www.nationalarchives.gov.uk/currency-converter/#currency-result. Accessed 22 September 2024.

19 Tanturri (1012), p. 81.

20 Gómez Beltrán (2022), p. 91.

21 Ibid., p. 93.

22 Corbett (1898), p. 171.

23 S.P. Dom. CCII. 14. In Corbett (1898), p. 142.

24 https://www.nationalarchives.gov.uk/currency-converter/#currency-result. Accessed 23 September 2024.

25 Gómez Beltrán (2022), p. 143. Four ships arrived in mid-May, sent by Drake with wounded and sick crew members. After the storm in early June, the *Golden Lion* deserted and returned to England alone. Drake, accompanied by four ships and the carrack *São Felipe*, arrived at the end of June. It is known that another eight ships arrived throughout August. As for the remaining seven, no documentation exists confirming their arrival.

26 Hutchinson (2013), p. 221.

27 Parker & Martin (2011).

28 For instance, *Archivo General Simancas, Guerra Antigua., Sección Mar y Tierra*, documents 205-334.

29 Nexon (2009), p. 223

30 Document dated 27 January 1586. Calendar of State Papers Foreign: Elizabeth, Volume 20, September 1585-May 1586. In: https://www.british-history.ac.uk/cal-state-papers/foreign/vol20/pp322-345. Accessed 14 August 2024.

31 Letter from the Queen to Leicester, dated 1 April 1586. Calendar of State Papers Foreign: Elizabeth, Volume 20, September 1585-May 1586. In: https://www.british-history.ac.uk/cal-state-papers/foreign/vol20/pp510-523. Accessed 14 August 2024.

32 Letter from Richard Cavendish to Burghley dated 10 April 1586. Calendar of State Papers Foreign: Elizabeth, Volume 20, September 1585-May 1586. In: https://www.british-history.ac.uk/cal-state-papers/foreign/vol20/pp523-540. Accessed 18 August 2024.

33 Letter from Leicester to the Privy Council dated 27 July 1587. Calendar of State Papers Foreign: Elizabeth, Volume 21, Part 1, 1586-1588. In: https://www.british-history.ac.uk/cal-state-papers/foreign/vol21/no3/pp197-217. Accessed 19 August 2024.

34 Van der Essen (1937), p. 191.

35 Letter from Philip II to the Duke of Parma, dated 1 April 1588. AGS, E-Castilla, docs. 165-174 and 175.

36 Document 4755 dated 2 April 1588. Letter from Hierónimo Lippomano, Ambassador of Venice in Spain, to the Dux and the Senate.

37   For instance, letter from the Duke of Parma to the King dated 20
      March 1588. Document 241. Calendar of State Papers, Spain (Simancas),
      Volume 4, 1587-1603. In: https://www.british-history.ac.uk/cal-state-
      papers/simancas/vol4/pp225-243. Accessed 22 August 2024.

38   Letter from Bernardino de Mendoza to Philip II dated 5 April 1588. AGS,
      E-Francia, doc. K-1567-53.

39   A good example of this can be found in Howard's letter to Walsingham
      dated 24 June 1588, where he states: 'It is very strange that at this point
      the delegates cannot discern whether the Spaniards seek a genuine
      peace or use the negotiation to buy time for other purposes'. It is
      important to note that both were staunch advocates for war against
      Spain (SP, Dom. Eliz., CCXI. 18). In a letter the following day, Howard
      assures Walsingham that the peace treaty is literally a 'deception' by
      Philip (SP, Dom. Eliz., CCXI. 26).

40   Van Der Essen, op. cit., p. 213.

41   Memorandum dated 20 June 1588. AGS, E-Flandes, leg. 594-101.

Chapter 4: The Enterprise of England

1   Juan de Zúñiga y Requeséns (1536–86), born into two of the most
      prominent noble families in Spain, held several high-trust positions
      during the reign of Philip II. These included serving as ambassador
      to Rome, Viceroy and Captain General of the Kingdom of Naples,
      and President of the Council of State. Among his most notable
      achievements was persuading Pope Gregory XIII to support Philip
      II's claim to the Portuguese throne. He is widely regarded as the
      King's principal advisor – prudent, incorruptible, and a figure of great
      authority.

2   Jensen (1988), p. 629.

3   Gómez Beltrán (2022), p. 258. Some authors, such as Parker (2008,
      p. 312), mistakenly attribute it to Juan de Zúñiga.

4   Bernardino de Escalante (1537–1605), originating from the lower nobility
      but well-connected to the court, joined the entourage accompanying
      Philip II to England for his marriage to Mary Tudor when he was
      under 20 years old. He quickly stood out for his remarkable intelligence,
      which earned him trusted positions at court. An excellent sailor and
      cosmographer, his works were highly influential in Europe, particularly
      his *Discurso de la navegación que los Portugueses hacen a los Reinos y
      Provincias de Oriente, y de la noticia que se tiene de las grandezas del Reino
      de la China* (Discourse of the navigation made by the Portuguese to the
      kingdoms and provinces of the Orient, and of the existing knowledge
      of the greatness of the Kingdom of China), published in 1577 and
      translated into English two years later.

5   Gómez Beltrán (2022), p. 215.

6   Sousa (1988), pp. 32–3, citing the Archivo dos Açores as a source, which unfortunately does not provide the names of the corsairs or the ships involved in the raid.

7   Gómez Beltrán (2022), pp. 149–50.

8   Hutchinson (2013), p. 52.

9   Gómez Beltrán (2022), p. 152.

10   General Archive of Simancas. Guerra Antigua. Sea and Land Section, bundles 2'1-230. D2858

11   This figure allows us to gauge the significance of Drake's capture of the carrack *San Felipe*, whose value amounted to 4 per cent of the total worth carried by the Spanish Treasure Fleet.

12   Gómez Beltrán (2022), p. 222.

13   Jensen (1988), p. 635.

14   Pedro Enríquez de Guzmán de Acevedo (1525–1610) was a distinguished military officer who spent much of his career serving in the Kingdom of Naples and the Duchy of Milan. After arriving in Lisbon, he took part in the city's defence during the Drake–Norris's expedition of 1589.

15   Gómez Beltrán (2022), p. 304.

16   Ibid., p. 312.

17   AGS, Secretariat of State for Fleets and Galleys, bundles 455-174 y 175.

18   On this point, an important clarification must be made. In the nineteenth century, the historian Cesáreo Fernández Duro erroneously transcribed the aforementioned letter from the duke, asserting that the duke became seasick while sailing – an assertion that is untrue. This error has led generations of historians to underestimate Medina Sidonia's capabilities to carry out his responsibilities in the Enterprise of England.

19   AGS, Secretariat of State for Fleets and Galleys, bundles 455-183 y 184.

20   AGS, Secretariat of State for Fleets and Galleys, bundles 455-226 y 227.

21   Gómez Beltrán (2022), p. 337.

22   Ibid., p. 333.

23   Pinzelli (2022), p. 111.

24   In the modern Var department in the Provence-Alpes-Côte d'Azur region in south-eastern France. He had been the first to scale the city walls. The defenders threw a stone at him, and he fell into the moat. Seriously wounded, he was taken to Nice, where he died a few days later. The Emperor Charles I, on hearing of the impending death of one of his best officers of the Tercios and a famous poet, had the entire garrison of Le Muy executed in reprisal.

25   Defined by Cervantes as 'The Phoenix of Wits' and 'Monster of Nature',
      he is considered the second-best author of the time after the author
      of Don Quixote. Among his works include: '*Fuenteovejuna*' (1619), '*El
      caballero de Olmedo*' (*The Knight from Olmedo*, 1620), '*La Dorotea*' (1632).
      He took part in the Battle of Terceira Island.

26   Pedro Calderón de la Barca is especially recognised for his 120 plays,
      among which the brilliant *Life Is a Dream* (*La vida es sueño*, 1629–35), *The
      Mayor of Zalamea* (*El Alcalde de Zalamea*, 1636), *The Siege of Breda* (*El sitio
      de Breda*, 1625), and *The House with Two Doors* (*Casa con dos puertas*, 1629)
      stand out, among many others. In his youth, he led a dissolute and
      excessive life, combining it with his literary work and serving as an
      officer in the Tercios. He eventually took holy orders, spending the final
      years of his life at court.

27   O'Donnell (1990), p. 29.

28   Jurado Riba (2022), p. 59.

29   We should not exclusively understand bandits as, following the
      Cambridge Dictionary: 'a thief with a weapon, especially one belonging
      to a group that attacks people travelling through the countryside'. In
      the Catalan case, these bandits were also connected to the private wars
      that Catalan nobles constantly declared against one another, which
      constituted an endemic problem within the Principality.

30   William Semple (1546–1633). Scottish nobleman and soldier. Following
      Mary Stuart's abdication, he went into exile in Flanders, where he
      entered the service of the Dutch rebels. In 1582, he switched allegiance
      to the Spanish side, serving under the Duke of Parma. He participated
      in the development of plans for the Enterprise of England and carried
      out diplomatic missions with King James of Scotland. In addition to
      continuing his involvement in several military campaigns in Flanders,
      he served as an advisor to three Spanish monarchs – Philip II, Philip III,
      and Philip IV – who regarded him as one of their foremost experts on
      British Isles politics.

31   William Stanley (1548–1630) was an English Catholic nobleman and
      soldier. His baptism of fire occurred in Flanders in 1567, during a
      period when Elizabeth I was still allied with Spain. Three years later, he
      was sent to fight against Irish rebels, remaining in Ireland until he was
      dispatched back to Flanders as part of Leicester's expedition, serving
      as one of Leicester's closest advisors. His intense hatred for Elizabeth
      I and James VI led him to conspire against them, though his attempts
      ultimately failed. He spent the rest of his life in the Low Countries,
      faithfully serving Spain.

32   O'Donnell (1990), p. 253.

33   Ibid., p. 248.

34   Carlo Spinelli de Castrovillari, Prince of Cariati, (1536–1609). He was
      Neapolitan.

35   Camillo Capizucchi, Marquis of Poggio Catino (1537–97). He came from
      one of the oldest and most prestigious families in Rome. His Italian
      Tercio was instrumental in the capture of Antwerp. Parma considered
      him one of his best generals, with whom he shared a close friendship.
      Shortly after Farnese's death, he requested permission to return
      to Rome. In 1597, he led a papal army to expel the Ottomans from
      Hungary, but during the retreat, he fell ill and passed away.

36   Marc de Rye de la Pallud, Marquis of Varenbon (1545–98).

37   Karl von Habsburg, Marquis of Breisgau (1540–90), a cousin of Philip II
      and a member of the Austrian branch of his family.

38   O'Donnell (1990), p. 77.

39   Van der Essen, op. cit., p. 217.

40   Gómez Bletrán (2022), p. 348.

41   Linés Escardó (1998), p. 80.

42   AGS leg. 221-189 y Doc. 6500, in González-Aller Hierro, Dueñas Fontán,
      et alt. (2014).

43   Doc. 6814, in González-Aller Hierro, Dueñas Fontán, et alt. (2014).

44   Docs. 5996, 6112 (sheet #2) y 6236.

45   According to Pedro Coco Calderón; meanwhile, Captain Calderón
      states that it was 10 o'clock in the morning.

46   We assume that, at the end of the sixteenth century, one ducat
      corresponded to 11 reals, and one pound sterling corresponded to
      33.47 reals. Thus, 50,000 ducats would correspond to £18,075. For the
      conversion from 1580 sterling to today's sterling, I follow the table in:
      https://www.nationalarchives.gov.uk/currency-converter/#currency-
      result.

47   Doc. 6154, in González-Aller Hierro, Dueñas Fontán, et alt. (2014).

48   State Paper, vol. II, pp. 101–02. *Mathew Starke's Deposition.*

49   García-Torralba (2021), p. 28.

50   Chinchilla (2023), p. 164.

51   Van der Essen, op. cit., p. 222.

52   Ibid., p. 224.

53   Calvar, Campo et alt. (2018), p. CXLII.

54   Bourne (1576), pp 7–8. The 8. Devise. 'If that an armie of Shippes doo
      ride in any Haven or Ri | ver to defend any place, or to keepe any place
      for recei | uing of any more strength or vittailes, & that they meane to
      ride there still, and have placed themselves in such order, that no ship
      may passe by them, either to man a place, or to vittaile that place, thus
      it may be done: first this, prepare such a sufficient number of bad or
      olde ships as shall be convenient, and then put such kind of stuffe into
      them as will quickly bee fiered, and then when you doo see convenient

time that the winde and tide dooth serve your turne, send those olde ships before with a fewe men for to governe them under saile, and with boates to save themselves, and then let them a little before set them on fire, and lay the principall ships aboord crosse their Halce or Stemme, and then there is no doubt but they shall drivv them to let slip their Anckers, or consume them with fire, and then the moe men that there bee in number aboord, the greater shalbe their terror, if that they have not boates enough to saue themselves, and then presently after that, you may come in, and doo your exployte, for that they will bee in such a maze with the fire, that you may doo what you list: for if this devise had been put in practise by Countie Mongomery when he went vnto Rochell, there is no doubt but they had both discomfited, or spoyled all those ships that did ride before Rochel, and also the Count Montgomery might have landed at his pleasure'. The Device 9 also expresses itself in very similar terms.

55   Chinchilla (2023), p. 120.

56   A detailed account of Moncada's final moments is provided by Fernández Duro (op. cit., p. 102), who cites the narrative written by one of Moncada's servants. In this final engagement, the owner of the galleass, the Barcelona merchant Joan Setantí i Desbosc, also lost his life.

57   González-Aller (2012), p. 21.

58   Ibid., p. 49.

59   Calvar, Campo et alt., op. cit., p. CLVII. Other authors, such as Álvaro Ocariz (2014, p. 43), claim that the captain was Ochoa de Goyaga.

60   Calvar, Campo et alt., op. cit., p. CLIV.

61   Martín and Parker, op. cit., p. 311.

62   In 1599, Van der Does had the opportunity to lead his own Dutch Armada against Spain – a little-known event that can only be briefly summarised here. His Armada consisted of seventy-four warships divided into three squadrons. However, the Dutch fleet suffered a catastrophic failure in its attacks on La Coruña, Cádiz, and the Canary Islands. Subsequently, the fleet split apart; half of the ships returned to the Dutch Republic, while Van der Does proceeded to attack Portuguese colonies along the West African coast. There, his crew was decimated by malaria, a disease that ultimately claimed his life as well.

63   Chinchilla, p. 168. The two English ships involved were part of the fleet under Dutch command that had blockaded the Duke of Parma's army. Consequently, they did not participate in the voyage through the English Channel.

64   Van der Essen, p. 225.

65   Gómez-Centurión (1987), p. 72.

66   O'Donnell, p. 402.

67   Calvar, Campo et al, p. CLVIII.

68   Van der Essen, p. 225.

69   Calvar, Campo et al, p. CLVII.

70   Doc. 6814, in González-Aller Hierro, Dueñas Fontán, et alt. (2014).

71   Hale (1911), p. 136.

72   Lambert (2011), p. 26.

73   State Papers, Vol. II, pp. 93–4.

74   Ibid., p. 96.

75   Ibid., p. 163.

76   Lambert, op. cit., p. 28.

77   State Papers, Vol. II, p. 183.

78   Martin/Parker, 1988, p 258.

79   Baltasar de Zúñiga y Velasco (1561–1622) was the younger brother of the Count of Monterrey and brother-in-law to the Count-Duke of Olivares, who supported his political and diplomatic career upon his return to Spain. He became a highly influential figure under King Philip III, serving as ambassador in Brussels, Paris, and Prague, as well as acting as Hofmeister to the future Philip IV.

80   Chinchilla, op. cit., p. 185.

81   Dos años más tarde, el obispo fue encarcelado y ejecutado por su ayuda a los españoles.

82   Este capitán de los Tercios de Alonso de la Ysla estaba al mando del barco, si bien el capitán era Juan Gregorio de la Peña.

83   Doc. 6500, in González-Aller Hierro, Dueñas Fontán, et alt. (2014).

84   AGS Archives, K. 1567. Doc. 6770.

85   Doc. 6706, in González-Aller Hierro, Dueñas Fontán, et alt. (2014).

86   Doc. 7127, in González-Aller Hierro, Dueñas Fontán, et alt. (2014).

87   Chinchilla, op. cit., p. 204.

88   The largest was the sinking of the *Castillo de Olite*, a troop transport ship, which during the Spanish Civil War was bombed by Republican troops, killing 1,476 Francoist soldiers.

89   Chinchilla, op. cit., pp. 244–5.

90   Ibid., p. 213.

91   Ritchie, Neil: 'Spanish Galleon of Tobermory Bay' in scottishhistory. org: https://www.scottishhistory.org/articles/tobermory-galleon/, accessed 14 March 2024.

92   Doc. 7141, in González-Aller Hierro, Dueñas Fontán, et alt. (2014).

93   Docs. 6999, 7045 y 7068, in González-Aller Hierro, Dueñas Fontán, et alt. (2014).

94   Porras Arboleda (2014), p. 78.

95   Doc. 6480, in González-Aller Hierro, Dueñas Fontán, et alt. (2014).

96   Doc. 6481, in González-Aller Hierro, Dueñas Fontán, et alt. (2014).

97   Calvar, Campo et alt., op. cit., p. CLXXIV.

98   Doc. 6621, in González-Aller Hierro, Dueñas Fontán, et alt. (2014).

99   A highly interesting figure, Joan de Cardona i Requesens (1530–1609) descended from two of the most prominent Catalan noble families of the Middle Ages and the Early Modern period. His uncle, Berenguer de Requesens, trained him in the art of warfare. He participated in the Disaster of Gelves (1560), the Siege of Malta (1565) – where he distinguished himself through heroic conduct – the Battle of Lepanto (1571), where he once again demonstrated exceptional bravery, and the conquest of Tunis (1573). He served as a royal advisor and was part of the working group tasked with reforming the Armada following the failure of 1588.

100   Calvar, Campo et al., op. cit., p. CLXXX. This detailed analysis of the ships, interestingly, coincides with the figures published in a book released in England in the autumn of 1588, titled *Certaine Advertisements out of Ireland, concerning the losses and distresses happened to the Spanish Navie, upon the West coastes of Ireland, in their voyage intended from the Northerne Isles beyond Scotland, towards Spaine*, printed in London by I. Vautrollier for Richard Field. This source reports a loss of 32 ships and 10,185 men from the Armada. The book is referenced in Document 7008. Consequently, the figures provided by historians such as Fernández Duro, Lewis, Padfield, or Martin and Parker, among others, would be inaccurate.

101   Calvar, Campo et alt., op. cit., p. CLXXXI.

102   Gómez-Aller (2012), p. 22.

103   Chinchilla (2023), p. 172.

104   Ibid., p. 179.

105   Ibid., p. 197.

106   Ibid.

107   López de Haro (1622), p. 499 and Chichilla, p. 228.

108   Chinchilla, p. 214.

109   Ibid.

110   Gómez-Aller (2012), p. 29.

111   Domínguez Núñez (2011), p. 228.

112   Chinchilla, op. cit., p. 291.

113   These amounts would correspond, respectively, to £283,218 and £257,470 when adjusted to 2017 values. For the sailors and soldiers, a

ransom equivalent to approximately £2,220 in today's currency (as of 2024) was paid.

114 By land, success had already been achieved following the Ottoman failure at the Siege of Vienna (1529).

115 It should be clarified that the Battle of Trafalgar was not, per se, the cause of the decline of the Spanish Empire. However, the severe economic crisis Spain was undergoing meant that the ships of the line that were lost could not be replaced – some of which, such as the *Santísima Trinidad* and the *San Juan de Nepomuceno*, were among the finest in the world – and those requiring repairs or maintenance were left to rot in the shipyards due to a lack of funds.

116 Gómez Beltrán (2013), p. 212.

117 Let us remember: the *María Juan*, the *San Mateo*, and the *San Felipe*.

118 Gómez Beltrán (2013), p. 212.

119 Ibid., p. 414.

Chapter 5: Anatomy of a Battle: Gravelines 1588

1   Davis (2001), p. 199.

2   Thucydides: *The Peloponnesian War*, 7.53.4: 'The rest the enemy tried to burn by means of an old merchantman which they filled with faggots and pine-wood, set on fire and let drift down the wind which blew full on the Athenians. The Athenians, however, alarmed for their ships, contrived means for stopping it and putting it out, and checking the flames and the nearer approach of the merchantman, thus escaped the danger', From London, J. M. Dent; New York, E. P. Dutton. 1910. Perseus Digital Library, accessed on 15 March 2024. A tactic that remained common throughout the Roman Empire and the Middle Ages.

3   In: https://www.tutorchase.com/answers/a-level/history/discuss-the-impact-of-the-spanish-armada-on-elizabeth-i-s-reign. Accessed on 15 March 2024.

4   In: https://www.gcsehistory.com/faq/spanisharmada.html, accessed on 15 March 2024.

5   Martínez González (2018–19).

6   Juan Escalante de Mendoza (1529–96). He was one of the finest sailors of his generation, Chief Pilot of the Casa de Contratación in Seville, and in 1595 was appointed Captain General of the Armada and Fleet of New Spain.

7   This work is divided into three books. The first was published in 1880, but the complete text was not published until 1985.

8   All of them had been successful admirals of the Indies Fleet, as well as renowned sailors, cosmographers, and military men.

9   Gómez Beltrán (2013), p. 300.

10   Kostman (2008).

11   Gómez Beltrán (2013), p. 242.

12   Martin and Parker (2002), p. 397.

13   Cristóbal Lechuga (1557–1622) He was a soldier, artilleryman and engineer, serving as Lieutenant General of Artillery in the States of Flanders and Milan, as well as a writer. He fought in Flanders, applying effective innovations during the sieges of Maastricht, Tournai, and Antwerp. Considered one of the finest artillerymen of his century, he authored several treatises that had a significant influence on the development of this branch of warfare. His most important work is *Discurso que trata de la Artillería y de todo lo necesario a ella, con un tratado de Fortificación y otros advertimientos* (1611). For details on rate of fire, see page 183 of this treatise.

14   Gómez Beltrán (2013), p. 277.

15   Ibid., pp. 206–09.

16   Ibid.

17   Doc. 7163. Fray Juan de Vitoria. *News from the English campaign of 1588*, in González-Aller Hierro, Dueñas Fontán, et alt. (2014).

18   Martin and Parker (2022), p. 396. It can be consulted at State Papers, Vol. II, pp. 258–9.

19   Gómez Beltrán (2013), p. 211.

20   Ibid., p. 548.

21   Raleigh (1826), p. 108.

22   Martin and Parker (2022), p. 397.

23   Doc. 6814. Captain Alonso Vanegas, *Events of the Armada from its departure from La Coruña to its return*, in González-Aller Hierro, Dueñas Fontán, et alt. (2014).

24   Gómez Beltrán (2013) p. 559.

25   McElvogue (2015), p. 133.

26   Gómez Beltrán (2013), p. 241.

27   Ibid., p. 238.

28   Interview in the newspaper *ABC* dated 16 September 2017, in: https://abcblogs.abc.es/espejo-de-navegantes/otros-temas/colin-martin-no-habra-jamas-nada-como-la-gran-armada.html, accessed 17 March 2024.

29   Gómez Beltrán (2013), p. 272.

30   Brimacombe (2000), p. 192.

31   Davis (2001), p. 202.

32   Decades later, Spain's enemy nations became aware of the existence of this treaty. In 1584, the Dutch seaman Lucas Janszoon Waghenaer managed to compile some fragments in the *Spieghel der zeevaerdt*, which was translated into English in 1588 as *Mariner's Mirror*. In Spain, the book was published for the first time in 1992. Today, the book has been digitised by the Royal Academy of History and has already been consulted over one million times.

33   The galleys, as the reader already knows, were only armed with artillery at the bow.

34   Gómez Beltrán (2013), p. 242.

35   Parker (1996), p. 94.

36   Martin and Parker (2022), p. 397.

37   Doc. 7122.2. Juan Gómez de Medina to Philip II. *Summary of his services and request for payment of what is owed to him by the Royal Treasury*, in González-Aller Hierro, Dueñas Fontán, et alt. (2014).

38   Ships that sailed alone or had fallen behind the protection of the Indies Fleet.

39   Nieva Sanz (2019), p. 86.

40   In https://www.bbc.co.uk/bitesize/guides/zq4s9qt/revision/5, accessed 17 March 2024.

41   Fagel (2003).

42   Thompson (1969), p. 215.

43   García Rivas (2009).

44   González Alonso (2014).

45   Fernández Secades (2008).

46   From this work, Admiral and historian Guillén Tato (1943), p. 9, cites fifteen editions in French, four in Dutch, three in Italian, and two in English.

47   The foreign policy of the Catholic Monarchs based much of its diplomatic activity on forging alliances through marriage. One of their main allies was Portugal. In fact, both Charles I and Philip II had Portuguese wives. The House of Medina Sidonia was no stranger to these alliances, arranging marriages with the House of Braganza from the early sixteenth century onwards.

48   Thompson (1969), p. 198.

49   Ibid., p. 213.

50   It should be noted that an important colony of English Catholic merchants had settled in Cádiz, having decided to establish themselves in Spain out of fear of Elizabethan repression.

51   Gómez Beltrán (2013), pp. 157–8.

52   Doc. 6601. Letter from Girolamo Lippomano, Ambassador of Venice in Spain, to the Dux and the Senate: *Llegada de la armada a las costas* ('Arrival of the Armada on the Coast').

53   Doc. 6739. Letter from Vincenzo Alamanni to the Cardinal Grand Duke of Tuscany: *Noticias de la Armada* ('News of the Armada').

54   Doc. 6538. Letter from Giulio Battaglino, Secretary of the Tuscan Embassy in Madrid, to the Grand Duke of Florence: *Report on what Baltasar de Zúñiga has recounted regarding the outcome of the Armada. Comments circulating at the Spanish Court.* Doc. 6628 – Letter from Vincenzo Alamanni to the Cardinal Grand Duke of Tuscany: *Opinions on the English Campaign.* Doc. 6750 – Letter from Ambrosio Spinola to the Cardinal Grand Duke of Tuscany: *News circulating regarding the outcome of the Armada and the King's intentions to pursue the Enterprise of England the following year.* Doc. 6779 – Letter from Giovanni Gritti, Venetian Ambassador in Rome, to the Dux and the Senate: *Justification of the Duke of Parma poorly received by Spanish ministers.* Doc. 7163 – Report by Friar Juan de Vitoria: *News regarding the English Campaign of 1588.* Doc. 6685 – Report by Fernando de Ayala: *Commentary on the events of the English Campaign of 1588.* Doc. 7163 – Report by Friar Juan de Vitoria: *News regarding the English Campaign of 1588.*

55   Doc. 6685 Report by Fernando de Ayala. *Comments on the events of the English campaign of the year 1588.* Doc. 7163. Report by Friar Juan de Vitoria. *News of the English campaign of 1588.*

Chapter 6: Beyond 1588

1   Parker (2014), p. 331.

2   Kamen (2003), p. 308.

3   Gorrochategui (2020), p. 53.

4   Wernham (1988), p. XIV.

5   Hermano de John Norris.

6   Gorrochategui (2020), pp. 62 and 63.

7   Hume (1896), p. 26 and note 17.

8   In the Mediterranean, the Battle of Lepanto (1571) and the Japanese invasion fleets over Korea in the Battles of Myeongnyang (1597) and Chilchonnyang (1597) involved a greater number of ships.

9   Gorrochategu, (2020), pp. 85–6.

10   Ibid., p. 116.

11   It must be considered that this entailed increased food consumption and that António of Crato had requested that Portuguese properties not be looted to avoid provoking popular resentment.

12   Brother of the Marquis of Santa Cruz, Álvaro de Bazán.

13   Apart from the two ships lost in La Coruña (and the three captured),
     some Anglo-Dutch vessels had already withdrawn from the
     expedition.

14   Gorrochategui (2020), p. 221 and Vaz and Falcao de Fonseca (1996),
     p. 49.

15   Gorrochategui (2020), p. 239.

16   Hume (1896), p. 71.

17   Wernham (1932), p. 236.

18   Hume (1896), p. 72.

19   Gorrochategui (2020), p. 269.

20   Ibid., p. 265.

21   Ibid., p. 280.

22   Originally, command of the fleet had fallen to Sir Walter Raleigh, but
     days into the voyage, his flagship was hit by a pinnace from the Queen
     with the order to return. Elizabeth had just learned that Raleigh had
     secretly married her Gentlewoman of the Privy Chamber, Elizabeth
     Throckmorton, without her permission, and they had already had a
     child. For this intolerable insult to his authority (and perhaps a fit of
     jealousy), the privateer spent a few months locked up in the Tower of
     London.

23   Wernham (1932), p. 173.

24   Williamson (1920), p. 239.

25   Monson (1703), p. 181.

26   Williamson (1920), pp. 126–32.

27   Ibid., pp. 71–2. Monson, en su relación de los hechos, Monson (1703), pp.
     179–80.

28   John Chudleigh (1564–89), like Drake, he hailed from Devon, but
     unlike the famous privateer, John was born into a noble family. In
     1583, he took part in Sir Humphrey Gilbert's first Arctic expedition. Of
     him, Markham says (1877, p. 19): 'He was of a right martial, bold, and
     adventurous spirit, and the famous actions of Drake and Cavendish
     ran so much in his mind, that he could not rest without undertaking
     to show himself the third Englishman that had circumnavigated the
     world and performed some noble service for his country'. The headland
     located on the eastern shore of Killiniq Island (Canada), is name Cape
     Chidley or Chudleigh after him.

29   Jowitt (2017), p. 9.

30   Hawkins (1747), s. 225.

31   The Gentlemen Aventurers: Richard Hawkins. In: http://
     www.gentlemenadventurers.org/index.php?option=com_
     content&view=article&id=136&Itemid=616. Accessed 6 October 2024.

32   Gorrochategui (2020), pp. 293–8, Fernández Duro (1897), pp. 103–15 and Rodríguez González (2001).

33   Monson (1703), p. 182.

34   Kamen (2003), p. 301.

35   Fitzgerald was the nephew of the famous Gerald FitzGerald, Earl of Desmond, who led the Desmond Rebellions in 1569–3 and 1579–83. Maurice had also been a member of the 1588 Armada.

36   Cantero Bonilla (2023), p. 85.

37   Simpson (2001), p. 102.

38   Fernández Duro (1897), p. 121. Quote: '40 well-armed ships, the rest hulks and trading vessels of 200 tons and under, and 80 boats for reconnaissance and landing. The Dutch contributed 20 of the naos, hulks and depot ships'

39   Cantero Bonilla (2023), p. 87

40   Parker (1996), p. 94. También en Richards (2003), Gorrochategui (2020), pp. 299–302. Fernández Duro (1897), pp. 121–7.

41   Cantero Bonilla (2023), p. 88.

42   Gorrochategui (2020), p. 303.

43   Cantero Bonilla (2023), p. 91.

44   Carlos Morales (2013).

45   Cantero Bonilla (2023), p. 97; Gorrochategui (2020), p. 304.

46   Fernández Duro (1897), p. 166.

47   Desde 1585, Catherine Michaela was married to Carlo Emanuele, duke of Savoy.

48   Fernández Duro (1897), p. 79.

49   But also a base for privateering, where the name of Pedro de Zubaiur stands out, who, with spectacular hand strikes, with only five filibotes, will become a real danger for the English merchant ships in the channel.

50   Isabel Clara Eugenia (1566–1633) was not only Philip II's favourite daughter, but she was also his most perceptive, cultured and shrewd one. She was Governor of the Low Countries between 1598 and 1633, carrying out an intelligent policy that came close to ending the war and where she was a patron of painters such as Rubens and Brueghel.

51   Pañeda Ruiz (2023), pp. 115–16 and Gorrochategui (2020), pp. 286–7.

52   Fernández Duro (1897), p. 90, quoting a French source.

53   The loss shocked England and, in particular, its queen, Mary Tudor, who left for history her famous quotation: 'When I am dead and opened, you shall find Philip [her husband] and Calais lying in my heart'.

54   Wernham (1932), p. 177.

55   Since the trade embargo imposed by Spain, the profitable commercial
network that English merchants had established with Andalusian
ports – where goods arriving from the Americas were traded and where,
paradoxically, they had enjoyed the friendship and protection of the
Captain General of Andalusia, the Duke of Medina Sidonia – effectively
disappeared. Furthermore, the outbreak of war also led to the closure
of ports to English merchants in various locations along the Flemish
and German coasts. To this must be added the increase in privateering
actions against English ships by Flemish, Spanish, and French forces.
Regarding the latter, there exists a particularly interesting report in the
Calendar of State Papers entitled *Depredations upon English ships by French
pirates from the year 1585, for which justice is demanded in France*, dated
July 1585. It can be consulted at: https://www.british-history.ac.uk/
cal-state-papers/foreign/vol21/no1/pp53-63.

56   Waida (2023), p. 468.

57   A sort of Prime Minister.

58   Irish (2018), p. 147.

59   Williamson (1920), pp. 227–30.

60   Originally known as Leicester House, the residence later referred to
as Essex House was constructed in 1575 by Sir Robert Dudley, Earl of
Leicester, on land that had formerly belonged to the Knights Templar
during the medieval period. In 1588, following Leicester's death, the
house was inherited by his stepson, Robert Devereux, after which it
became known as Essex House. The building was demolished around
1670.

61   In Ireland, Essex had knighted numerous officers who were blindly
loyal to him and who now accompanied him in his plans to overthrow
the government.

62   Cecil Papers: June 1601, 1–15. In: https://www.british-history.ac.uk/
cal-cecil-papers/vol11/pp214-233. Accessed 11 October 2024.

63   Haigh (2000), p. 170.

64   Catherine Howard (1550–1601) was the granddaughter of Mary Boleyn
and great-niece of Anne Boleyn. In 1563, she married Charles Howard,
Lord Howard of Effingham, Lord High Admiral of England and the
first Earl of Nottingham. In 1572, she was appointed First Lady of the
Bedchamber to Elizabeth I, becoming one of the Queen's closest friends
and confidantes, whom she served with loyalty for 45 years.

65   She was the daughter of Lord Effingham and Catherine Howard and
had been widowed a few years earlier following the death of Sir Robert
Southwell.

66   Herrero Sánchez (1993), pp. 181–2. Salt was an excellent preservative
for products such as fish or cheese, for instance. The exploitation
of the Araya salt flats proved especially productive, offering salt of

exceptionally high quality and easy accessibility. As Herrero Sánchez has noted, during that period over 100 Dutch ships per year arrived at the Araya Peninsula.

67   It was a conspiracy aimed at overthrowing James I and placing his cousin, Lady Arabella Stuart, on the throne. In any case, Raleigh's involvement was deemed to have been secondary.

68   Salas Amela (2021), p. 43.

Chapter 7: A Rainy Day in Valladolid: the Treaty of London

1   Haigh (2000), p. 171.

2   Ibid., p. 175.

3   Ibid., p. 170.

4   Letter from Philip III to Albert of Austria, dated 15 May 1603. AGS, State, leg. 2224/2, doc. 442.

5   Rasilla Hidalgo (2016).

6   Tassis became the first Spanish ambassador to England since the expulsion of Bernardo de Mendoza in 1584. After negotiations, the ambassador became Pedro de Zúñiga.

7   Martínez Ruiz (2023), p. 117.

8   Ibid..

9   Hutchings and Cano-Echevarría (2012), p. 92.

10   In English: A General Collection of Treatys (Vol. 2). In: https:// books.google.es/books?id=U8nYFSTQhXcC&pg=PA131&redir_ esc=y#v=onepage&q&f=false. Accessed 17 December 2024, pages 131-146; and in Spanish and Latin in *Colección de los tratados de paz, alianza, neutralidad, garantía...* In: https://books. google.es/books?id=YdrziOQCoK8C&pg=PA243&redir_ esc=y#v=onepage&q&f=false. Accessed 17 December 2024, pages 243–69.

11   Parker (2008), p. 335.

12   Martínez Ruiz (2023), p 119.

13   Sobre su accidentado viaje, Williams (2009), p. 32.

14   Pérez Gil (2004), p. 6.

15   Williams (2009), p. 40.

16   Hutchings and Cano-Echevarría (2012), p. 96. They recount the incident in which Lerma, a malevolent figure, humiliated Howard during the ceremony, and Howard's subsequent expression of his desire to return immediately to England. However, the King intervened to alter Howard's decision.

17   Ibid.; and Williams (2009).

Conclusion

1   At the risk of being repetitive, history offers numerous examples of truly decisive naval victories. In Antiquity, for instance, we find the Battle of Salamis (480 BC), in which a coalition of Greek city-states defeated the Persian Empire's fleet, resulting in the loss of approximately 50 per cent of its ships; or the Battle of Actium (31 BC), where the fleet of Mark Antony and Cleopatra suffered a similar percentage of losses against Augustus's forces (more precisely, Agrippa's). Beyond the aforementioned early modern examples, the sixteenth and seventeenth centuries also witnessed significant naval outcomes. In Asia, during the Imjin War, the Battle of Noryang (1598) saw the Japanese fleet lose nearly half of its vessels to the Korean navy under the legendary Admiral Yi Sun-sin. And one need not look far to find a truly decisive defeat for the Spanish Armada: in 1639, the Battle of the Downs resulted in the loss of nearly 30 per cent of the fleet commanded by Antonio de Oquendo (son of Miguel de Oquendo). Spain's inability to recover from this blow in the short term consolidated Dutch naval dominance in their territorial waters and contributed to Spain's eventual recognition of Dutch independence through a negotiated peace shortly thereafter. Later military history would provide further paradigmatic examples of decisive naval victories, including Trafalgar (1805) and Midway (1942), though further examples are beyond the present scope.

2   Mattingly (2005), p. 397.

3   Simpson (2001), p. 102.

4   A battle that brought together several descendants of key figures from the Anglo-Spanish War of 1585–1604, such as Edward Cecil, 1st Viscount Wimbledon (1572–1638), grandson of William Cecil, Baron Burghley; Robert Devereux, 3rd Earl of Essex (1591–1646), son of the Elizabethan Devereux; and Manuel Pérez de Guzmán (1579–1636), son of Alonso Pérez de Guzmán.

5   The original quote by Durant is: '"The two greatest problems in history", says a brilliant scholar of our time, are "how to account for the rise of Rome, and how to account for her fall". We may come nearer to understanding them if we remember that the fall of Rome, like her rise, had not one cause but many, and was not an event but a process spread over 300 years. Some nations have not lasted as long as Rome fell'. It appears in the epilogue of his work *The Story of Civilization: Caesar and Christ*, on page 665 of the Simon and Schuster edition, New Tork, 1944.

6   Haigh (2000), p. 170.

7   Ibid., p. 171.

8   Shakespeare wrote a total of ten plays about English kings: *King John, Richard II, Henry IV* (two parts), *Henry V, Henry VI* (three parts), *Richard III*, and *Henry VIII*. It is thought that *Richard II* may implicitly be a criticism of the Queen's weakness and lack of a successor.

9   Knerr (1989), p. 7.

10   The Tudor Trilogy is a group of three operas composed by Gaetano Donizetti that revolve around the figure of Elizabeth I of England; these are: *Anna Bolena,* premiered on 26 December 1830, at the Teatro Carcano, Milan; *Maria Stuarda,* on 30 December 1835, at the Teatro alla Scala, Milan; and *Roberto Devereux,* on 28 October 1837, at the Teatro San Carlo, Naples. Verdi's *Don Carlo* was first performed on 11 March 1867 at Le Peletier in Paris.

11   Who was neither King nor was his name Arthur.

12   The Frame Semantics Theory by George Lakoff explains how people understand the world through mental structures called frames. These cognitive frames shape how individuals interpret language, ideas, and experiences. They are deeply embedded and often unconscious, guiding reasoning and decision-making. In politics, but also in Arts, controlling the frame can influence public opinion, as the same facts can be perceived differently depending on how they are framed. Thus, language is never neutral, as it activates frames that shape meaning and reflect broader worldviews and values.

13   Walker (2004), p. 124.

14   Corbett (1917), p. 176.

BIBLIOGRAPHY

Archives

Archivo General de Simancas.

Museu Marítim de Barcelona.

Calendar of State Papers: 1547-1649: Tudor and Stuart government papers.
In: https://www.nationalarchives.gov.uk

Documents

ABREU Y BERTODANO, José Antonio, *Colección de los tratados de paz, alianza, neutralidad, garantía . . . hechos por los pueblos, reyes y príncipes de España con los pueblos, reyes, príncipes, republicas y demás potencias de Europa . . . : desde antes del establecimiento de la monarchia gothica hasta el feliz reynado del Rey N.S. D. Phelipe V.* Published by Diego Peralta, Antonio Marín and Juan de Zuñiga. In: https://books.google.es/books?id=YdrziOQCoK8C&pg=PA243&redir_esc=y#v=onepage&q&f=false

CABRERA, Rodrigo de la, *Relacion muy cierta y verdadera, que trata de la iornada que el Serenissimo Principe Cardenal Don Alberto de Austria, que por mandado de su Magestad, fue a entender en las cosas de los estados de Flandes, sobre la toma y sucesso de Cales*, Sevilla, 1596.

GONZÁLEZ-ALLER HIERRO, José Ignacio, DUEÑAS FONTÁN, Marcelino de, CALVAR GROSS, Jorge and MÉRIDA VALVERDE, M.ª del Campo, *La Batalla del Mar Océano*, Ministerio de Defensa. Armada Española. Volumes 1, II, III and IV. Madrid, 2014.

KNAPTON, J. J.; and KNAPTON, P., *A General Collection of Treatys*, Volume 2. In: https://books.google.es/books?id=U8nYFSTQhXcC&pg=PA131&redir_esc=y#v=onepage&q&f=false

LECHUGA, Cristóbal, *Discurso del capitán Cristóbal Lechuga: en que trata de la artillería, y de todo lo necesario à ella, con un tratado de fortificación y otros advertimientos . . .*, en el Palacio Real y Ducal, por Marco Tulio Malatesta. Milán, 1611.

PAZ SALAS, Pedro de, *La felicissima armada que el rey Don Felipe nuestro Señor mandó juntar enel puerto de la ciudad de Lisboa enel Reyno de Portugal / hecha por Pedro de Paz Salas.* - Em Lixboa: por Antonio Aluarez, 9 Mayo 1588. - [22] f.; 2°

SIMÓN, Fray Pedro: *Noticias historiales de las conquistas de Tierra Firme en las Indias occidentales* (1565); published by M. Rivas, Bogotá (1882). https://archive.org/details/tierrafirmeindias01simbrich/tierrafirmeindias01simbrich/

VAZQUEZ, Alonso, *Los sucesos de Flandes y Francia del tiempo de Alejandro Farnese.* Imp. de Miguel Ginesta. Madrid, 1879 *The Works of Sir Walter Raleigh,* vol I. Burt Franklin: Research and Source Works Series #73, New York, 1829.

Books

ALBERRO, Solange (2015), *Inquisición y Sociedad en México, 1571-1700,* Fondo de Cultura Económica. México.

ÁLVARO OCÁRIZ (2016), José Andrés, 'Presencia vasca en la Armada española (III)' in *Revista de Historia Naval,* n° 135, pp. 29–44.

AMAYA PALACIOS, Sebastián (2021), 'La estatalización de las Armadas castellanas en el Caribe (siglo XVI)' in *Revista de Historia Naval,* n. 152, pp. 7–30.

BALBI, Francesco (2005), *The Siege of Malta, 1565,* Boydell Press.

BEN SRHIR, Khalid and BIN AL-ṢAGHĪR, Khālid (2005), *Britain and Morocco During the Embassy of John Drummond Hay, 1845-1886,* Psychology Press.

BLANCO NÚÑEZ, Jose María (2023), 'La Contrarmada (1589)' in *Guerra Anglo-Española 1585-1604,* LXVII Jornadas de Historia Marítima. Instituto de Historia y Cultura naval, Madrid.

BRIMACOMBE, Peter (2000), *All the Queen's Men: the World of Elizabeth I,* St. Martin's Press. New York.

BURRAGE, Henry S. (ed.) (1906), *Early English and French Voyages, Chiefly from Hakluyt, 1534-1608,* Charles Scribner's Sons, pp 135–48. www.americanjourneys.org/aj-031/. Accessed 8 February 2024.

CARLOS MORALES, Carlos Javier (2013), 'Endeudamiento dinástico y crisis financieras en tiempo de los Austrias: las suspensiones de pagos de 1557-1627' in *Librosdelacorte.es,* n° 7, año 5 Otoño-invierno.

CASAS, B. (2015), 'Piratas y corsarios en el Golfo de México: la presencia de John Hawkins en San Juan de Ulúa (1568)', *Antropología. Revista Interdisciplinaria Del INAH* (100), pp. 51–66. In: https://revistas.inah.gob.mx/index.php/antropologia/article/view/8207, accessed 8 February 2024

CASTRIES, Henry de (1918), *Sources Inédites de l'Histoire Du Maroc, France, Pays-Bas, Angleterre.* Vol. I, Luzac & Co.

CHINCHILLA, Pedro L. (2023), *Los prisioneros de La Armada Invencible: La historia nunca contada sobre los capturados de la gran armada española de 1588.* Ediciones B. Madrid.

CORBETT, Julian Stafford (1899), *Papers relating to the Navy during the Spanish war, 1585-1587*, Navy Records Society.

CORBETT, Julian Stafford (1917): *Drake and the Tudor Navy*. Longman's, 1917.

DAVIS, Paul K. (2001), *100 Decisive Battles: From Ancient Times to the Present*, Oxford University Press.

FARIA, Vicomte de (1917), *Descendance de D. Antonio, Prieur de Crato*, Imprimeries Réunies, Lausanne.

FERNÁNDEZ DURO, Cesáreo (1897), *Armada Española: desde la unión de los Reinos de Castilla y León* (Vol. 3). Sucesores de Rivadeneyra. Madrid.

FRÉNÉE, Samantha (2015), 'Pirates and Gallows at Execution Dock: Nautical Justice in Early Modern England' in *Actes du colloque: Les Fourches Patibulaires du Moyen Âge à l'Époque moderne. Approche interdisciplinaire*. In: https://doi.org/10.4000/criminocorpus.3080. Accessed 23 July 2024.

GARCÍA HERNÁN, Enrique (1999), *La cuestión irlandesa en la política internacional de Felipe II*, Tesis Doctoral. Universidad Complutense de Madrid.

GAJDA, Alexandra (2023), 'War, peace and commerce and the Treaty of London (1604)', *Historical Research*, Volume 96, Issue 274, pp. 459–72, https://doi.org/10.1093/hisres/htad011. Accessed 10 October 2024.

GOLDINGHAM, C. S.: *The Expedition to Portugal, 1589*. https://doi.org/10.1080/03071841809421877. Accessed 20 December 2024

GÓMEZ BELTRÁN, Antonio Luis (2013), *La Invencible y su leyenda Negra*, Arin 2013 Ediciones, S.L. Benalmádena.

GÓMEZ BELTRÁN, Antonio Luis (2017), *Islas Terceiras. La Batalla Naval de San Miguel*, Ediciones Platea, 2017.

GONZÁLEZ-ALLER HIERRO, José Ignacio (2012), 'El combate de Isla Flores (8 de septiembre de 1591)' in *Revista de Historia Naval*, Year XXX, n° 116, pp. 9–12.

GORROCHATEGUI, Luis (2020), *Contra Armada*. Editorial Crítica. Barcelona.

GUILLÉN TATO, Julio F. (1943), *Europa aprendió a navegar en libros españoles*, Instituto Histórico de Marina. Barcelona.

GUTIÉRREZ NÚÑEZ, Francisco Javier (2011), 'Los Solís Manrique (siglos XVI-XIX). Señores de Ojén y marqueses de Rianzuela' in *Takurunna*, year 1, pp. 217–72.

HAIGH, Christopher (2000), *Elizabeth I*, Longman.

HAKLUYT, Richard (1907), *The Principal Navigations, Voyages, Traffiques & Discoveries of the English Nation: Made by Sea Or Overland to the Remote & Farthest Distant Quarters of the Earth at Any Time Within the Compasse of These 1600 Yeares*. Published by J.M. Dent & Sons. In: https://www.perseus.tufts.edu/hopper/text?doc=Perseus%3atext%3a1999.03.0070. Accessed 23 May 2024.

HAWKINS, Richard (1747), *The Observations of Sir Richard Hawkins, Knt in his Voyage into the South Sea in the Year 1593*. Editor: Charles Ramsay Drinkwater Bethune. The Hakluyt Society, London.

HUME, Martin A. S. (1896), *The Year After the Armada and Other Historical Studies*. Macmillan and Co., New York.

HUTCHINGS, Mark and CANO-ECHEVARRÍA, Berta (2012), 'Between Courts: Female Masquers and Anglo-Spanish Diplomacy, 1603–5' in *Early Theatre* 15.1.

HUTCHINSON, Robert (2013), *The Spanish Armada*, Hachette UK.

IRISH, Bradley J. (2018), 'The Dreading, Dreadful Earl of Essex', in *Emotion in the Tudor Court: Literature, History, and Early Modern Feeling*, Northwestern University Press, pp. 137–78. JSTOR, https://doi.org/10.2307/j.ctv3znz47.9. Accessed 11 October 2024.

ITA RUBIO, Lourdes de (2017), 'Extranjería, protestantismo e Inquisición: presencia inglesa y francesa durante el establecimiento formal de la Inquisición en Nueva España' in *Signos Históricos*. Vol. XIX, n° 38, pp. 8–55.

JAMES, Alan (2012), 'A French Armada? The Azores Campaigns, 1580-1583', *The Historical Journal*, vol. 55, no. 1, pp. 1–20. JSTOR, http://www.jstor.org/stable/41349643. Accessed 15 July 2024.

JENSEN, De Lamar (1988), 'The Spanish Armada: The Worst-Kept Secret in Europe', *The Sixteenth Century Journal*, vol. 19, no. 4, pp. 621–41. JSTOR, https://doi.org/10.2307/2540990. Accessed 25 September 2024.

JONES, Susan (2022), 'In Truth, They Are My Masters: The Domestic Threat of Early Modern Piracy', *Humanities*, 11: 123, 2022. https://doi.org/10.3390/h11050123. Accessed 23 July 2024.

JOWITT, Claire (2017), "The Hero and the Sea: Sea Captains and Their Discontents" in *L'Empire*, number 74. https://doi.org/10.4000/1718.888, accessed 8 July 2024

JURADO RIBA, Víctor (2022), 'La elección de generales en la flota de Tierra Firme (1584): los méritos mediterráneos de don Lluís de Queralt para la dirección de una flota atlántica', *El taller de la Historia*, 14(1), pp. 45–66. DOI: https://doi.org/10.32997/2382-4794-vol.14-num.1-2022-4017. Accessed 2 October 2024.

KAMEN, Henry (2003), *Empire: how Spain became a world power, 1492-1763*. HarperCollins.

KNERR, Douglas (1989), "Through the "Golden Mist": a Brief Overview of Armada Historiography" in *American Neptune*, 49 (1), 5-13.

LAVIÑA, Javier, MENDIZÁBAL, Tomás, PIQUERAS, Ricardo ITZEL DE GRACIA, Guillermina, HIDALGO PÉREZ, Marta; TOUS, Meritxell, LÓPEZ, Rubén and TRESSERRAS JUAN, Jordi (2015), 'Don Luis de Mozambique, el que elegido fue de su rebelión por rey primero: Santiago del Príncipe, primer pueblo de negros libres de América' in *Informes y trabajos. Excavaciones en el exterior*, N°. 12, pp. 247–58,.

LAMBERT, Andrew (2011), *Admirals*, Faber & Faber. London.

LINDEN, H. Vander (1916): 'Alexander VI. and the Demarcation of the Maritime and Colonial Domains of Spain and Portugal, 1493-1494' in *The American Historical Review*, Volume 22. In: https://archive.org/details/jstor-1836192/page/n1/mode/2up. Accessed 21 January 2024.

LINÉS ESCARDÓ, Alberto (1984), 'Las condiciones meteorológicas durante la navegación de la Real Armada de Lisboa a Coruña' in *Revista de Historia Naval*, Año n° 2, N° 4, pp. 67–74.

LINÉS ESCARDÓ, Alberto (1988), 'Las desfavorables condiciones meteorológicas que precedieron al viaje de la gran Armada contra Inglaterra' in *Nimbus*, numbers 1 and 2, pp. 79–84.

LUCENA SALMORAL, Manuel (1992), *Piratas, Bucaneros, Filibusteros y Corsarios en América*, Eed. Mapfre. Madrid, 1992.

MACKIE, J. Duncan (19260, 'James VI. and I. and the Peace with Spain, 1604', *The Scottish Historical Review*, vol. 23, no. 92, pp. 241–9. JSTOR: http://www.jstor.org/stable/25525604. Accessed 10 October 2024.

MARTIN, Colin & PARKER, Geoffrey (2011), *Armada: The Spanish Enterprise and England Deliverance in 1588*. Yale University Press.

MacLEAN, Gerald and MATAR, Nabil (2011), *Britain and the Islamic World, 1558-1713*. Oxford University Press.

McELVOGUE, Douglas (2015), *Tudor Warship Mary Rose*. Bloomsbury Publishing.

McBRIDE, Gordon K. (1973), 'Elizabethan Foreign Policy in Microcosm: The Portuguese Pretender, 1580-89' in *Albion: A Quarterly Journal Concerned with British Studies*, Vol. 5, No. 3, pp. 193–210.

MARKHAM, Sir Clements Robert (1877), *The Voyages of Sir James Lancaster, Kt., to the East Indies*, B. Franklin. New York.

MARTIN, Colin (1978), 'La Trinidad Valencera. A Spanish Armada Wreck' in *Archaeology*, Vol. 31, No. 1, pp. 38–47. https://www.jstor.org/stable/41726856, accessed 2 February 2024.

MARTÍNEZ GONZÁLEZ, Alfredo José (2018–19), 'Memoriales presentados a Felipe II de Cristóbal de Barros y Diego Marroquín sobre fomento naval, comercio, legislación e instituciones cantábricas (1570 aprox.)' in *AHDE*, tomo LXXXVIII-LXXXIX. In: https://www.boe.es/biblioteca_juridica/anuarios_derecho/abrir_pdf.php?id=ANU-H-2018-10055500592. Accessed 1 May 2024.

MARTÍNEZ RUIZ, Enrique (2023), 'El cambio de coyuntura y la paz entre España e Inglaterra (1595-1604)' in *Guerra Anglo-Española. 1585-1604*. LXVII Jornadas de Historia Marítima. Ciclo De Conferencias, Cuaderno Monográfico N.° 87. Instituto de Historia y Cultura Naval, Madrid, pp. 101–21.

MATTINGLY, Garrett (1990), *The Defeat of the Spanish Armada*, Penguin.

MESA GALLEGO, Eduardo de (2021), 'Intervenciones militares inglesas en las guerras de las monarquías hispánicas (1500-1660')' in TAULER CID, Benito (coord.), *Presencia británica en la milicia española / The British Presence in the Spanish Military*, Revista Internacional de Historia Militar N. 99. Madrid, 2021.

MONSON, William (1703), *Sir William Monson's Naval Tracts*, A. and J. Churchill, 1703. In Biblioteca Nacional de Portugal. DOI: https://purl.pt/30261. Accessed 4 October 2024.

NÚÑEZ JIMÉNEZ, Carlos (1994), 'Linaje de los Guzmanes' in *Aljaranda: revista de estudios tarifeños*, N°. 14, 1994, pp. 22–7.

O'DONNELL Y DUQUE DE ESTRADA, Hugo (1990), *La fuerza de desembarco de la Gran Armada contra Inglaterra (1588)*, Editorial Naval.

PAÑEDA RUIZ, José Manuel (2023), *Crónica de un fracaso anunciado: la campaña bretona de Felipe II*, UNED.

PARKER, Geoffrey (1996 i), *The Military Revolution: military innovation and the rise of the West, 1500–1800*, Cambridge University Press.

PARKER, Geoffrey (1996 ii), '¿Por qué triunfó el asalto a Cádiz en 1596?' in BUSTOS 1996. – Cádiz: Servicio de Publicaciones de la Universidad: Fundación Municipal de Cultura del Ayuntamiento, p. 93–124.

PARKER, Geoffrey (2014), *Imprudent King: a new life of Philip II*, Yale University Press.

PÉREZ GIL, Javier (2004), 'Visitas reales a Astorga en el siglo XVIII según las fuentes inglesas' in *Argutorio*, n°14. Astorga.

PINZELLI, Eric G.L. (2022), *Masters of Warfare: Fifty Underrated Military Commanders from Classical Antiquity to the Cold War*, Pen and Sword.

PORRAS ARBOLEDA, Pedro Andrés (2014), 'La aportación de Castro Urdiales ala Armada Invencible (1586-1618) in *Cuadernos de Historia del Derecho*, n°11, pp. 53–111. http://dx.doi.org/10.5209/rev_CUHD.2014.v21.47718. Accessed 14 March 2024.

QUINN, David B. (1979), 'England and the Azores, 1581-1582: Three Letters' in *Revista da Universidade da Coimbra*, vol XXVII, UC Biblioteca Geral 1., Coimbra, pp. 205–17.

RASILLA HIDALGO, Carlos (2016), *Las Conversaciones de Boulogne: el Fracaso de la Diplomacia Anglo-Hispánica entre las Paces de Vervins y Londres*, Facultad de Filosofía y Letras. Universidad de Cantabria. Santander.

ROBINETT, Boatswain C. M. (1952), 'Ship Technology And The Defeat Of The Armada' in *Proceedings*, Vol. 78/2/588, US Naval Institute. https://www.usni.org/magazines/proceedings/1952/february/ship-technology-and-defeat-armada, Accessed 10 June 2024.

RODRI□GUEZ, Manuel (Dir.): *El asalto anglo-holandés a Cádiz en 1596 y su contexto internacional.* Conferencias del Curso de Verano de la Universidad de Cádiz, 13-19 de Julio de RODRÍGUEZ GARAT, Juan (2023), 'En la Tercera el francés' in *Revista De Historia Naval*, number 162, pp. 11–54. https://doi.org/10.55553/603sjp16201. Accessed 22–24 July 2024.

RODRÍGUEZ GONZÁLEZ, Agustín Ramón (2001), 'La última expedición de Drake y Hawkins' in *Historias de la Mar* in *Revista General de Marina* (240-1), 79–84.

RODRÍGUEZ GONZÁLEZ, Agustín Ramón (2011), *Drake y la 'Invencible'*, Ed. Sekotia, Madrid.

RONCIÈRE, Charles de la (1899), *Histoire de la marine française*. Volume 4, E. Plon, Nourrit, Paris.

RULE, Margaret (1981), 'The Raising of the Mary Rose', *Journal of the Royal Society of Arts*, vol. 129, no. 5298, pp. 332–43. JSTOR, http://www.jstor.org/stable/41373302. Accessed 29 July 2024.

RUMEU DE ARMAS, Antonio (1947), *Los viajes de John Hawkins a América (1562-1595)*. Publicaciones de la Escuela de Estudios Hispano-Americanos de Sevilla. Serie Monografías n°9, Sevilla.

SANTAMARTA LOZANO, Luz María (2001), *Don Guerau de Spes en la corte isabelina la documentación diplomática y el conflicto anglo español en la segunda mitad del siglo XVI (1568-1571)*, Universidad de Oviedo.

SEGAS, Lise (2017), 'Cimarrones y corsarios: de la realidad colonial a la épica histórica' in *Hipogrifo. Revista de literatura y cultura del Siglo de Oro*, vol. 5, núm. 2, pp. 241–60.

SEVILLA GONZÁLEZ, María del Carmen (2016), 'Las nupcias de Catalina de Aragón. Aspectos jurídicos, políticos y diplomáticos' in *Anuario de Historia del Derecho Español*, vol. LXXXVI, Ministerio de Justicia. Madrid.

SIMPSON, William (2001), *The Reign of Elizabeth*, Heinmann Advanced.

SOLÍS DE LOS SANTOS, José (2009), 'Relaciones de sucesos de Inglaterra en el reinado de Carlos V' in *Testigo del tiempo, memoria del universo: cultura escrita y sociedad en el mundo ibérico (siglos XV-XVIII)*. Coord. Manuel Fernández, Carlos-Alberto González-Sánchez, Natalia Maillard Álvarez, pp. 640–98.

SOUSA, Nestor De (1988), 'Sinais de presença britânica na vida açoreana (séculos XVI-XIX)'. *Arquipélago. História*. Número Especial, pp. 25–100. In: http://hdl.handle.net/10400.3/1058. Accessed 23 September 2024.

SUÁREZ BILBAO, Fernando (2020), *La historia en la pantalla: hechos, propaganda y manipulación. El caso de Felipe II*. Dykinson.

THOMPSON, I. A. A (1969), 'The Appointment of the Duke of Medina Sidonia to the Command of the Spanish Armada' in *The Historical Journal*, vol. 12, no. 2, pp. 197–216. JSTOR, http://www.jstor.org/stable/2637801. Accessed 11 April 2024.

THOMPSON, I. A. A. (1975), 'Spanish Armada Guns' in *The Mariner's Mirror*, Volume 61, Issue 4, pp. 355–71. In: https://doi.org/10.1080/00253359.1975.10658046, accessed 27 March 2024

VAN DER ESSEN, H. (1959), *Etat de la 'maison' de Marguerite de Parme, gouvernante des Pays-Bas, 1560-1566. Analyse de documents inedits des archives farnesiennes*, Bulletin de la Commission Royale d'Histoire.

VAZ, Joao (1996), and FALCAO DE FONSECA, Luis, 'Sir Francis Drake and the 'poor king' D. Antonio: The Portugal Voyage of 1589' in *The British Historical Society of Portugal*, 36. Lisbon.

VIGIL MONTES, Néstor (2021), 'La copia de tratados diplomáticos en el archivo regio inglés por parte de la primera embajada permanente de los reyes católicos en Inglaterra (1487-1508)' in *Estudios de Historia Moderna* from Ediciones Universidad de Salamanca. N. 43, n. 2, pp. 39–70. DOI: https://doi.org/10.14201/shhmo20214323970

WALKER, Julia (2014), *The Elizabeth Icon: 1603-2003*. Palgrave Macmillan.

WALSH, Micheline (1979),'The Military Order of Saint Patrick, 1593' in *Seanchas Ardmhacha: Journal of the Armagh Diocesan Historical Society*, vol. 9, no. 2, pp. 274–85. JSTOR, https://doi.org/10.2307/29740927. Accessed 8 October 2024.

WERNHAM, R. B. (1932), 'Queen Elizabeth and the Siege of Rouen, 1591' in *Transactions of the Royal Historical Society*, vol. 15, pp. 163–79. JSTOR, https://doi.org/10.2307/3678646. Accessed 5 October 2024.

WERNHAM, R. B, (1988), *The Expedition of Sir John Norris and Sir Francis Drake to Spain and Portugal, 1589*, Aldershot: Temple Smith; distributed by Gower Publishing Co., Brookfield, Vt.

WILLIAMS, Patrick (2009), 'El Duque de Lerma y el nacimiento de la corte barroca en España: Valladolid, verano de 1605' in *Estudios de Historia Moderna*, 31, Esdiciones Universidad de Salamanca, pp. 19–51.

WILLIAMSON, James A. (1922), *A short history of British expansion*. Macmillan and Co., limited.

WOODHEAD, Christine (2009), 'England, the Ottomans and the Barbary Coast in the Late Sixteenth Century' *State Papers Online 1509–1714*, Cengage Learning EMEA Ltd.

INDEX